CLEAN AIR

All proceeds from this book will go to the following three charities:

Friends of the Earth https://friendsoftheearth.uk

The Environmental Law Foundation https://elflaw.org

Hebburn Helps https://hebburnhelps.co.uk

ISBN 978-1-7391024-0-1 (paperback) | ISBN 978-1-7391024-1-8 (hardback)

ISBN 978-1-7391024-2-5 (ebook) | ISBN 978-1-7391024-4-9 (ebook Amazon)

ISBN 978-1-7391024-3-2 (audiobook)

Edited by Lesley Wyldbore
Typeset by Latte Goldstein
Cover design by Paul Johnstone
Map by Najla Kay
Monkton Coke Works motif by Sofia Strass

To Jennie, the heroine of this book,

*I hope that this is a fitting tribute to you and
the members of the Hebburn Residents'
Action Group who battled alongside you.*

CONTENTS

Foreword

I was Director of Friends of the Earth when the extraordinary campaign that Jennie Shearan led was at full stretch.

CLEAN AIR provides a wonderful record of that campaign, and of the vision, courage, and staying power of Jennie and her close colleagues in the Hebburn Residents' Action Group. The notion of 'environmental justice' has always resonated very powerfully with me: there can be no justice for the Earth without justice for its people – and vice versa. But words about environmental justice are easy: the practice is so, so much harder. Jennie Shearan embodied that concept in every regard, providing inspiration and hope for countless people – including grizzled old campaigners such as myself!

As a Patron of the Environmental Law Foundation, it was so moving to be reminded of the crucial importance of the campaign against Monkton Coke Works in the establishment of the Foundation. These legacies need to be honoured – with the deepest respect and love.

Jonathon Porritt CBE

———

Jennie was a remarkable woman who was rooted in – and utterly devoted to – her local community, the people she felt honoured to represent as a member of Tyne and Wear County Council and their first and only woman chair.

For my part, I remember Jennie fondly and with huge respect from the time that I was elected member of the European Parliament for Tyne and Wear in 1979 and first made her acquaintance. I admired her courage and persistence in standing up for the people she represented and for being undaunted in pursuing her goal of tackling the unacceptable levels of pollution to which her neighbourhood was being exposed.

Pursuing a goal single-mindedly is not always popular and I know that persistence in pursuit of a cause can sometimes be unfairly characterised as obsessiveness, but after a long career in politics I have come to rate both persistence and commitment as essential qualities to achievement. In politics it is all too easy to get sidetracked and even to get overwhelmed by the sheer range of issues and causes to consider. It takes courage and staying power to pursue a cause through rebuffs and setbacks, and Jennie showed both of these qualities throughout to a remarkable degree.

Jennie's story is part of Tyneside's history and heritage. Furthermore, at a time when tackling climate change and environmental pollution is vital to secure the future of our planet, it is a fascinating case study of some of the challenges environmental campaigners find themselves confronting. Finally, at a time too when we celebrate the undoubted progress women have made in politics, this book is a timely reminder of the importance of honouring women such as Jennie who, although no longer with us, can still inspire others to follow her example of pursuing worthwhile causes and of giving selfless public service.

Baroness Joyce Quin, former MEP, MP and Europe Minister

Although it is over thirty years since the Monkton Coke Works inquiry, it remains the most traumatic in my career as a planning consultant. This is the story of an indomitable woman, Jennie Shearan, fighting for the health and well-being of a poor community in Newcastle against the combined power and influence of big business, supine regulators, and government.

With the help of three professionals who had only met shortly before a major inquiry, Jennie succeeded in accomplishing her vision for clean air in her community. It is an inspiring story about an exceptional woman, and it is also an economic and political history of Britain from the Jarrow Hunger Marchers in the 1930s to the legal implications of Brexit. This book is a must-read!

Dr Wendy Le-Las

——

MAP

The below map aims to help the reader immediately situate themselves into the heart of the story. It has been intentionally drawn using pencil, which is made from carbon, the main component in coal, a rock that has left an indelible mark on the area.

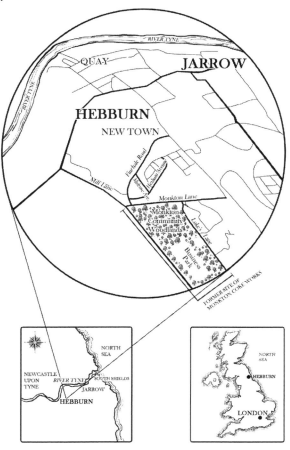

PART ONE

PARADISE LOST

CHAPTER ONE

HAW'WAY THE LADS

What is a miner – what is he worth?
When he spends his life in the bowels of the earth,
Where darkness surrounds him,
The air full of dust,
Work is no pleasure – hard feelings are just.
But who sympathises? – Who seems to care?
The bosses don't worry as long he's there.[1]

The people we become, the lives we lead, and the values we hold, owe much to the lottery of our birthplace.

Jennie Shearan was born at home in Hebburn, in the North East of England, on 26 November 1922, shortly before the Great Depression. The tenth of eleven children, her parents were so poor that Jennie never even had a teddy bear.

The North East of England comprises three urban areas, the largest of which is Tyneside, where Hebburn is situated. The region has brief mild summers followed by long cold winters. These seasonal changes contribute to a diverse landscape that includes extensive moorlands containing rare species of plants and animals and a beautiful coastline that has led to its classification as an Area of Outstanding Natural Beauty.

The people of Tyneside are affectionately known as Geordies, and Jennie was as proud a Geordie as they come. It is difficult to hear their inimitable accent without smiling, such is their famously friendly demeanour and the unique slang of their dialect. One of the theories around the genesis of the term Geordie is that it originated from the North East coal mines of the 1800s because the local miners, also known as pitmen, wore Geordie safety lamps, designed by George 'Geordie' Stephenson. It is, in many ways, a suitable explanation given that the area has such a strong heritage of industrial activity. The definitive example of a Geordie phrase is 'Haw'way the Lads'.[2] Today, a chant of encouragement from local fans during football matches to inspire the men to keep playing well, the origin of the phrase stems from mining. The pitmen would be called from different points in the mine, for example, 'all the way lads' or 'half-way lads', to help navigate their progress.

On the south bank of the River Tyne, the iconic river that flows through Tyneside, is the town of Hebburn. A predominantly residential area of some 16,000 inhabitants, it was initially a small fishing hamlet during the Anglo-Saxon period, belonging to the Church until the Dissolution of the Monasteries. In the 1600s, the Ellisons, a wealthy Newcastle family of merchant adventurers, acquired the land. They saw its commercial potential and in 1618 began extracting coal from the ground. This would inaugurate a long-standing symbiotic relationship between Hebburn and coal that would play an integral role throughout Jennie's life.

Coal is a combustible organic rock that forms when dead plant matter submerged in swamp environments is subjected to the geological forces of heat and pressure over millions of years. Over time, the plant matter transforms from moist, low-carbon peat, into an energy-dense carbon-rich dark sedimentary rock. This fossil fuel exists in underground formations called coal seams, and is extracted from the ground through mining.

The 1700s saw the UK transition to many new manufacturing processes at a time in history known as the Industrial Revolution. These processes not only radically impacted every aspect of people's daily life but enabled the country to become the world's leading commercial nation, controlling a global trading and military empire. Many factors contributed to rapid industrialisation. For example, high levels of agricultural productivity provided excess manpower and food, and the numerous ports, rivers, canals, and roads facilitated the cheap movement of materials. The national legal system in place also promoted business in a way that was advantageous relative to other countries. The North East benefitted from possessing all these features. However, the abundance of natural resources such as coal had the most significant impact on the area. The region sits on a coal seam that stretches from the northernmost part of Northumberland to the southernmost part of Durham.

'Coal was the king of the Industrial Revolution',[3] powering trains, making iron and steel, generating electricity, and fuelling engines. It quickly established itself as the backbone of the economy's advancement, and the North East became the central coal-producing area of the country. Due to the coal's relative accessibility via bell pits and its proximity to navigable rivers and wagonways, it could be efficiently mined and safely transported to the biggest market, London.

The coal owners of Northumberland and Durham thus established dominance of the industry, which continued throughout the 1700s thanks to the invention of Newcomen's atmospheric engine,[4] bringing previously unworkable areas into production. Deep shaft mining began to develop extensively in the late eighteenth century, further propelling the Industrial Revolution. The North East continued to play a critical role. The region was the site of the first deep pits in the country, which went as deep as 1,000 feet – the height of the Eiffel Tower.

There were significant perils for the miners venturing deep into the ground. They battled against low light, poor air circulation, and explosive gases. Conditions were dirty and cramped for the underground workforce, with explosions, cave-ins, poisonous gases, and falling rocks a possibility at any time.

Miners were also subject to various health stressors, ranging from hearing difficulties to lung damage. Some ninety per cent of miners were left with impaired hearing by the age of 50, and coal workers' pneumoconiosis is still a common illness today. Black lung disease, as it is more often referred to, occurs when coal dust settles on a miner's lungs after prolonged exposure. The disease takes many years to develop and often leads to impairment of the lungs, disability, and premature death.

Despite the significant drawbacks of such a dangerous occupation, coal mining offered numerous benefits to the pitmen and their families. It gave the men regular work and a steady income.

Large communities were built around collieries, and there was a strong sense of camaraderie among the miners. Each miner had a fellow 'marra'[5] who accompanied them underground as they tried to keep each other safe beneath the low ceilings. Courage and physical strength were required for the long hours of continuous hard labour. The miners risked their lives together, and this fostered a high degree of solidarity among the workforce. Mining became more than a job for the men working down the pit and the families they provided for. It became part of their identity.

Far removed from the pitmen who earned only a modest living were the coal owners making lucrative gains from the mines. In 1726, a group of the North East's most powerful coal owners formed 'The Grand Alliance', an agreement dated to run for 99 years. It included George Bowes, a successful coal proprietor and politician who sat in Parliament, as one of its signatories. The intent of the treaty that he oversaw was essentially to take over the coal industry in the North East, with the cartel aiming to prevent the opening of new collieries by buying up much of the available land. They proposed blocking off any coal property they could not directly access. It was an aggressive policy and it worked.

This coalition was to become a forerunner for an even more ambitious and monopolistic agreement known as 'The Limitation of the Vend', signed in 1771, which aimed to secure North Eastern coal to be priced just under the amount for which coal from other areas could be sold.[6] The price fixing depended on all the major coal owners adhering to the agreement, strict enforcement of output control, and ongoing prioritisation of the London market. The Grand Allies all complied and were thus able to maintain their pre-eminence for the following decades, delivering the largest share of the 11 million tonnes of coal produced nationally towards the end of the century.

As the Ellisons already owned land in Tyneside and had been mining for many decades before 'The Grand Alliance', the family were well positioned to open a large-scale coal operation. In 1792, they opened Hebburn Colliery, which eventually operated three coal pits. They were some of the first pits in the country to use the new method of tubbing, with wooden segments lining the shafts to stem water flow. Hebburn soon became a centre of innovation in the coal mining industry. The first mechanisation of pits in the north of England came when steam engines pumped water from Hebburn Colliery, and the owner's mansion, Ellison Hall, was the venue where the inventor Sir Humphry Davy, using bottled methane from a Hebburn mining pit, successfully tested his eponymous safety lamp.[7]

As a new century approached, the ongoing development of railways and locomotives enabled coal to travel faster to the factories consuming the fuel. Moreover, because the locomotives were also powered by coal, demand for coal rose even further. However, Parliament began to take a serious view of the amount of hazardous smoke and cinder emitted from the locomotive engines and the issues the pollution was causing for the people, animals, and wildlife near the line. The unflagging growth in coal consumption during the Industrial Revolution also caused unprecedented air pollution in industrial centres.

When combusted, coal produces harmful waste and noxious fumes, including sulphur dioxide, carbon dioxide, nitrogen oxide, sulphuric acid, ammonia, and arsenic. Long-term exposure to sulphur dioxide causes constriction of the lung airways. High levels of carbon dioxide and nitrogen oxides can affect respiratory function and cause chronic lung disease. Sulphuric acid is a highly corrosive chemical that can cause severe skin burns, induce breathing difficulties if inhaled, and burn holes in the stomach if swallowed. Breathing in high levels of ammonia and arsenic can cause a sore throat and irritated lungs, and exposure to high enough amounts can be fatal. Soot is another by-product of coal combustion, contributing to coronary artery disease, lung cancer, and other respiratory illness such as asthma to those exposed to it for extended periods. Acid rain is another atmospheric by-product, whereby compounds such as sulphur dioxide and nitrogen oxides mix with water and oxygen in the air. This type of rain has detrimental effects on trees and soils, destroys insects and aquatic life-forms, causes paint to peel, corrodes steel structures, and is unsurprisingly highly damaging to human health.

The oxymoronic fossil fuel that underpinned so many decisions in the country was yet to be considered as destructive as it was constructive. Across the early part of the nineteenth century, pollution regulation was weak. With the government intensely focused on establishing the country's hegemony on the world stage, there was little concern about coal's impact on the environment. However, as pollution continued to worsen across the country, measures needed to be implemented.

In 1825, the Liverpool and Manchester Railway Bill was introduced, with a clause specifying that coke had to henceforth be used to power the locomotives instead of coal. The production of coke requires a specific high carbon type of coal, called bituminous coal. When bituminous coal is heated to an extremely high temperature for up to 36 hours, the by-product of this thermal distillation is coke. Coke is considered a much cleaner fuel because it produces intense heat but little or no smoke when burned, contrary to coal.

The institutionalisation of coke as the nation's combustible of choice proved to be a great stimulus to the North East coalfields. The coal seams in the region were well known for producing bituminous coal. This positioned the North East owners as unchallenged leaders in a rapidly expanding market for the smokeless fuel. With the boom in locomotives also came a higher demand for iron and steel. These essential materials were needed for bridges and rail tracks that were springing up throughout the country and required coke for smelting.

By the 1830s, established landowners in the North East, alongside an influx of individuals with no experience in the coal industry, were racing to invest in old and new collieries. There were fortunes to be made in this coal and coke rush, and it was the elites who owned the land that stood to gain the most.

A descendant and beneficiary of one of the original signatories of 'The Grand Alliance' was John Bowes. His father was John Lyon-Bowes, Earl of Strathmore and Kinghorne, and his aunt was Elizabeth Angela Marguerite Bowes-Lyon. This was one of the most prestigious families in the country. In subsequent years, Elizabeth would go on to marry Prince Albert, Duke of York, and they would soon welcome their first child. The child, affectionately nicknamed Lilibet by her family, would later become Queen Elizabeth II. John Bowes and his relatives were at the highest level of the aristocracy. They *were* The Establishment.

Educated at Eton and the University of Cambridge, Bowes had little interest in the assets that he had inherited, which included castles and country estates spanning the north of England and Scotland. He preferred instead to pursue his interests in art and horse racing. However, by the mid-1830s, and with the onset of the coal and coke boom, Bowes began to look more closely at the land and wealth under his stewardship. He set about creating one of the great firms of the British coal industry, Bowes and Partners Limited.

Powered by more than 5,000 coal miners working tirelessly underground with menial pay, at its height, the firm was producing two million tonnes of coal and 250,000 tonnes of coke annually from its fifteen collieries across the North East.[8] By the early part of the twentieth century, Bowes and Partners had formed a successful partnership with the other great North Eastern heavy industry firm, Palmers Shipbuilding and Iron Company. Such was the power and influence of these well-connected business leaders that Palmers even built a ship named after the mine owner. The 'John Bowes' was the world's first purpose-built steam-powered cargo ship to carry coal.

Palmers was based in Jarrow, a neighbouring town to Hebburn. Similarly situated on the south bank of the River Tyne, Jarrow had also benefitted from the discovery of coal at the beginning of the Industrial Revolution. Unfortunately, by 1851, the easier seams of coal had been exhausted, and the Jarrow pits had become less profitable, leading to their closure. Thankfully, Palmers was established on Jarrow's riverbank in the same year. It soon became the largest shipbuilding centre in the country, supplying many of the world's navies and employing thousands of men. Palmers Hebburn was later established a mile down the river to provide ship repair services. The parent company became central to the area's economy, both for the numbers employed there and for the ancillary businesses that served the needs of the shipyard and town. The otherworldly Blondins cableway cranes, deployed to move heavy loads across the site, were unique to Palmers in the UK and became a symbol of the firm's potency.[9] From constructing steamships for the Italians to bridges for the Indians, the company prospered for decades, taking contracts from all over the world.

Two years after Palmers was formed, another shipbuilding company was established. In 1853, Andrew Leslie founded his eponymous firm in Hebburn and went on to produce everything from iron sailing ships and warships to cargo liners and paddle steamers. In addition, Leslie helped develop Hebburn into a modern town by providing housing, apprenticeships, education, and spaces to host religious and social events for his workers and their families. Other industries soon followed, with the establishment of brick, salt, metal, and chemical works.

Within a timespan of less than 200 years, Hebburn had grown from a small fishing hamlet into a modern town of several thousand inhabitants. Many successful firms went on to congregate around the elbow of the River Tyne in Hebburn. At the turn of the century, the French electrical engineer Reyrolles moved his scientific instruments plant from London to the riverbank. His self-named firm became a significant employer across Tyneside. Nearby was Tharsis Copper Works, Tennants Chemical Works, Jarrow Steel Works, and the cable producer Pyro, with the complementary industry leaders operating close to each other. The world-renowned Swan Hunters built ships on the other side of Hebburn Quay and assembled some of the most outstanding vessels in the world, including RMS *Mauretania*, which held the Blue Riband for the fastest crossing of the Atlantic, and RMS *Carpathia*, which rescued survivors from the Titanic. For centuries, from the dawn of the Industrial Revolution and moving into the early part of the twentieth century, Tyneside was a conurbation whose economy was proudly associated with heavy industry.

Unfortunately, World War I would have a devastating effect on the lives and economies of many nations around the world. By 1918, when the Allies declared victory, more than eight million soldiers had died worldwide, and several dynasties had been destroyed. Moreover, the prolonged, brutal, and expensive conflict had severely weakened the UK's economy. Additionally, the US and Japan had replaced the UK as leaders in international exports and lending. This would soon have a disastrous impact on the country's key export industries, including the shipbuilding firms in Tyneside.

It was in this tumultuous climate of economic turmoil that Jennie Shearan was born on a winter's day in 1922. She was the tenth of eleven children and was christened Eliza Jane but would become known as Jennie. Her father was a union shop steward and worked as hard as possible to provide for his large family. To help make ends meet, Jennie's mother would spend hours walking miles to the quayside to buy fish and then sell them from her basket to passers-by on her way home. Raising so many children was challenging in such difficult economic conditions.

The difficult way of life endured by Jennie and her family was common at this time and would continue to worsen throughout the 1920s. In 1925, still recovering from the effects of the war, the British Government took measures to restore the pound sterling to the gold standard at its pre-war exchange rate to stabilise the economy. While this succeeded in making the pound convertible to its intrinsic value in gold, the decision immediately made British exports more expensive on world markets because the price of gold had been overestimated by ten per cent.[10] With coal and steel becoming less competitive, global demand slowed drastically, and these industries cut costs by lowering workers' wages.

Another major event to exacerbate the economic decline was the 1929 Wall Street Crash in the United States. As world trade contracted, the British Government introduced trade tariffs, which further diminished demand for British products coming out of industrial areas. Staple export industries such as coal mining, shipbuilding, and steel were heavily concentrated in specific areas of the UK, including the North East, and the region suffered a great deal.

The Great Depression had begun. By the end of 1930, national unemployment had risen to twenty per cent of the country's workforce, and exports had fallen in value by fifty per cent.[11] With no unemployment benefits in place at the time, this led to the impoverishment of much of the UK's population, particularly in industrial areas. The North East remained depressed for most of the decade. Coal mining and shipbuilding continued to slowly decline from their Victorian heyday, becoming outdated, smaller, and less efficient. Many people experienced poverty, unemployment, and deprivation as the deterioration of these industries contributed to pockets of long-term unemployment, with some workers having no work for years. In some towns in the North East, unemployment reached as high as seventy per cent, and a government report in the mid-1930s estimated that around twenty-five per cent of the UK's population existed on a subsistence diet, often with signs of child malnutrition, such as scurvy, rickets, and tuberculosis.

In 1931, when Jennie was just nine years old, her mother passed away. It was a period of immense sadness for the whole family, and especially painful for Jennie and her young siblings. From this moment on, Jennie would no longer have the special love and guidance that only a mother can give. She developed a close bond with her father in the years that followed. He would often hold meetings in the front room of the family home for his fellow union members. As a young girl, Jennie would watch the men debate the day's issues and listen to street-corner preachers declaiming against social injustices. She would recount the elders' dialogue to her adolescent friends and discuss how these challenges could be addressed. A focal point of their conversations was how the community could tackle the acute long-term economic downturn that the area was enduring.

Palmers, which only a few decades earlier had been a thriving icon of British dominance in the Industrial Revolution, began incurring heavy losses. By 1932, the North East firm was insolvent, and two years later, dismantled. The community of Jarrow was devastated by the demise of their biggest employer. The mood in the town was desperate. Jarrow had become 'a workhouse without walls'.[12]

At the time, Jarrow's incumbent political party was Labour, and they had selected Ellen Wilkinson as its local parliamentary candidate for the next general election. Known as 'Red Ellen', owing to both the Socialist Party colour that she represented, as well as her flaming red hair, she was one of the first women in Parliament. Wilkinson embodied a sympathetic approach to politics that aimed to advance the position of the working classes.

She felt a strong personal connection with the people of Jarrow and the loss of the shipyard, which had become the life source for the community. Wilkinson saw a town of skilled workers who had no opportunity to work, and she represented their plight vociferously in Parliament. Around this time, an American entrepreneur, Vosper Salt, convinced that the world demand for steel was about to rise, sought to purchase the site where Palmers shipyard had stood.

However, he was confronted with the British Iron and Steel Federation. This organisation had ostensibly been formed by the government to rationalise the iron and steel industry, but in reality, it existed to protect it from outside competition. The Federation ultimately withheld the capital needed for the development of Salt's scheme, and the President of the Board of Trade told Parliament that 'Jarrow must work out its own salvation'.[13] This was 'the last straw in official cruelty'[14] in the eyes of Mrs Wilkinson. The president's callous comments 'kindled the town'[15] and inspired it to action.

Since the turn of the century, the National Unemployed Workers' Movement had organised regular hunger marches in which unemployed workers converged on London to confront Parliament, believing that this would improve conditions. Upholding this approach, it was decided that 200 of Jarrow's dislocated men would walk the 300 miles to London in October 1936, arriving at the start of the new parliamentary session the following month to present a petition requesting the re-establishment of industry to the House of Commons. The 26-day Jarrow March secured funding from all the local political parties, the town's churches, and the business community. It became a metaphor for the underdog's fight against anonymity. As Wilkinson had stated, 'Jarrow's plight is not a local problem ... it is the symptom of a national evil',[16] and the whole town pulled together in support. One marcher explained, 'We were more or less missionaries of the distressed areas, not just Jarrow'.[17] Their desperation was perhaps best summarised when one of the rain-battered marchers was seen packing the ham from a sandwich he had been given into an envelope to send back home for his wife and children, who had not eaten meat in weeks. In November, Mrs Wilkinson presented the 12,000 signatures to Parliament, stating, 'His Majesty's Government and this honourable House should realise the urgent need that work should be provided for the town without further delay'.[18]

Sadly, the government was largely indifferent, their response devoid of any meaningful connection to the men's predicament. Despite having won the hearts and minds of many across the country, the marchers were not granted a meeting with the Conservative Prime Minister, Stanley Baldwin. Wilkinson was grief-stricken by Parliament's abandonment of her constituents and was seen sobbing broken-heartedly in a quiet street near Westminster.

Nevertheless, while the global demand for British coal and shipbuilding had declined, the national need for coke had grown exponentially, both for domestic use and in steel furnaces. The government may have had little concern for the employment needs of the marchers, but they did have a coke quota that had to be met. The government subsequently authorised the construction of a coke works to be built in Hebburn, the adjoining town where the marchers were based.

The town of Hebburn can be divided into three parts.

First, the quay area, which starts at the level of the River Tyne and runs up to St Aloysius Church. This area had flourished during the glory days of the nearby Palmers shipyard and was where the majority of the Hebburn populace lived. However, following the company's demise, the area struggled, and much of the housing was earmarked for demolition. By the mid-1930s, Jennie's father had left the quay area and joined his older daughters in Liverpool on the west side of the country. As Jennie was still a teenager, she reluctantly accompanied her father. Liverpool has many similarities to her hometown: both are found in the north of England, the inhabitants speak with unique accents, and both have an emblematic river running through them. But, while Jennie felt comforted being with her siblings in a place that was not too dissimilar from her birthplace, she became sorely homesick.

The second part of Hebburn is the centre of the town, today widely known as the New Town, where many housing estates were later built and where locals visit its shopping centre and enjoy fishing and picnicking at the nearby Riverside Park.

Finally, there is the southern part of Hebburn, which spans the breadth of Monkton Lane and is only two miles from the town centre. The Monkton area was named after the monks who had originally mined coal from the nearby coal seam and consisted mainly of agricultural land, with little to no residential development.

Towards the end of 1936, following the government's orders to construct a coke works, the local council prepared a development plan. The planning department of the council allocated a twenty-acre section of woodland on the southern periphery of Hebburn for industrial purposes. At the time, the jurisdiction of the land fell under Durham County Council. Each local council had its own Local Planning Authority that was empowered to make urban planning decisions for the area. Following the government's authorisation, Hebburn's Urban District Council granted Bowes and Partners a permit to build the coking plant on the woodland.

Due to Monkton Lane's proximity to the Jarrow marchers' homes, and with most of the land there being agronomy-based greenfield, it was decided that the area would be suitable to construct the facility. The plant's purpose of producing smokeless coal was an environmentally and economically noble one, as it would help the country tackle its emerging smog and air pollution problem, while helping address the employment issue in the area. At a cost of £250,000, several million pounds in today's money, and taking a year to complete, Bowes and Partners laid down a battery of 33 silica coke ovens, a by-products plant, and a coal washery.

The industrial complex was ready for operation in the early part of 1937.[19] The battery, flare stack, gas holder, and purifiers occupied the centre of the site on a north-east to south-west axis formed by the Bowes Railway line, which crossed the complex. This wagonway delivered the coal into the facility. Coal would be stocked and blended prior to carbonisation in the north-western part of the site. Metallurgical coke, the finished product from the carbonisation process, would be stored in the south-east area, to be supplied to the steel industry for use in blast furnaces. The regenerative circulation ovens, the pusher, and the quenching tower standing on opposite ends of the plant, alongside the associated boilers and chimneys, combined to inextricably alter the area's skyline.

Work began at the site. Two hundred men were employed to operate the plant and manage the demanding and filthy process. First, the coal was placed in a crusher and mixed with water and oil. A larry car would then deliver the pulverised coal into the oven's charging port to be heated. Once a leveller arm had smoothed the pile of coal, each charging port was sealed by a process called luting. This would involve pouring a wet clay mixture around the edges of each charging port to prevent leaks from the charging port lids. The ovens were then sealed in a twenty-hour process, where the coke was heated by gas flames at over $1,000^0$C. This meant that the ovens would stay active for most of the day.

Once the coal had been fired in the vast ovens at high temperature, it would be taken to the cooling towers of water, generating dark carcinogenic clouds. The final step of thermal distillation, called pushing, involved using a ram to guide the molten coke out of the oven into a railroad car. This car transported the coke to a nearby quench tower, where the coke was showered with water to prevent the coke from igniting when exposed to open air. Extreme noise and volatile plumes of grit-infused steam would travel in whatever way the wind was blowing as the hot coke was quenched. If the coal had not been wholly carbonised during the pushing process, it would create a 'dirty push', with significantly more plumes of toxic black gases being thrust into the atmosphere from the quencher. The quenched coke would then be crushed into half-inch pieces, known as coke breeze, to be used domestically and industrially in blast furnace operations at steel and iron mills.

In emergencies, excess gas would need to be burned off, creating vast flames that would reach high into the sky. Such was the power of the flames that they would illuminate not just the whole plant but also the whole of south Hebburn. The production of coke also spawned the nauseating stench of rotten eggs.

Thankfully, given its remote location, the facility could perform its intense and clamorous task without directly polluting or disturbing anyone nearby with its unpleasant by-products of black dust and foul-smelling sulphurous fumes.

Despite this, in October of that year, Jarrow Council complained to Bowes and Partners about the fumes from Monkton Coke Works. Notwithstanding its relative isolation, the nefarious pollution from the site was affecting nearby people, animals, and plants. Looked at through a broader environmental lens, it was clear that the problems that burning coal caused were not being reduced, so much as being transferred from the area of consumption to the area of coke production, with anything nearby suffering the consequences. The firm replied that efforts would be made to eliminate the fumes, and it hoped soon to overcome the nuisance. Nonetheless, no significant measures were put in place. Monkton Coke Works operated sedulously throughout the 1930s and the onset of World War II in 1939 did nothing to halt its production levels. The plant continued to process 200,000 tonnes of coal a year, yielding ever-increasing tonnes of coke, through a dirty and cacophonous process that was mercifully far away from any housing.

CHAPTER TWO

HOME SWEET HOME

'Mrs Shearan, can you come over? Me mam's having a baby.'

World War II was to become the biggest and deadliest war in history, with the UK one of the many countries involved in the fight against the Nazis. Civilians at home had to endure the constant threat of bombing raids and gas attacks, and underground air raid shelters were constructed to protect communities. It was during one such bomb attack in Liverpool that Jennie met a Welsh sailor called David. The destroyer ship on which he was sailing had docked nearby and they had both sought refuge in the same communal shelter. She was by now twenty years old, five foot seven inches in height, with brown wavy shoulder-length hair framing her soft features and pale blue eyes. He was handsome, athletic, and blue-eyed. They soon fell in love and stayed in touch with letters when David went back to war.

During those years, while David was on a convoy through the Arctic Circle to maintain supplies to the Russians fighting the German invasion, Jennie worked at a munitions factory and faithfully waited for him.

While working at the factory, Jennie's fellow workers had asked her to speak to management on their behalf to secure better working conditions and a possible pay rise. Heavily influenced by her father, Jennie had developed a strong understanding of right and wrong and was willing to fight for what she believed in. She had an innate sense of what justice meant, never just for herself but for those around her. She spoke with the manager on behalf of her colleagues. When the manager had reflected on her request, he returned to Jennie with the news that the workers would receive their pay rise, but her services would no longer be required. This event did not deter Jennie but rather cemented her need to always try and do the right thing. Nevertheless, she was frightened to go home and break the news to her father, who knew how hard jobs were to come by. His response surprised Jennie. Although disappointed, he was also understanding, and told her 'you're a chip off the old block' when he heard the news.

In 1944, during a short break in David's naval duties during the war, he returned to Liverpool and the young couple married. It was a humble ceremony at a registry office, with no flowers or a wedding dress for the bride. The newlyweds honeymooned in David's mountainous hometown of Cymmer. When they missed their train on their way to meet David's family, the local miners, who had just finished their shift and were covered in coal dust, graciously allowed them on their bus and sang traditional Welsh mining songs together for the rest of their journey.

Jennie quickly fell pregnant with their first child, Barbara. As she was now married, Jennie made the decision to return to Hebburn. For the remainder of World War II, she raised her child without her absent husband, as did most women in the country at the time. With limited funds, Jennie lived with her sister and brother-in-law and their own two young children in their small home.

By the end of 1945, Hebburn, like every other British town, was recovering from the war. Major cities were bombsites throughout the country, with over one million homes destroyed or badly damaged. There was an acute shortage of decent housing, and every village, town, and city was financially and emotionally exhausted.

A daily reminder of the austerity that every citizen had to brave was ongoing rationing. Food rationing had been introduced after the Nazis had attacked many of the ships that brought supplies to the UK. Before the war, the country imported 55 million tonnes of food, but a month after the war started, this figure had plummeted to just 12 million. Clothes rationing was also in place, with the government limiting people to buying just one new set of clothes once a year. Even Queen Elizabeth II had to purchase her wedding dress with ration coupons.

The 'make do and mend' slogans that had circulated throughout the war, stressing the need to stop waste and unnecessary consumption, were still fresh in the public psyche. Recycling every conceivable material, and economising in every possible way, were everyday activities. Allotments were regarded by their owners as a key boost for the country's food production. The rag-and-bone man, with his horse and cart, was a regular visitor to neighbourhoods, collecting any unwanted household items in exchange for a sweetie for the children or a few shillings for the parents. If nature called for the horse while the junk dealer was collecting items, then someone nearby would, without fail, shovel up the excrement for future use as compost in the gardens. Absolutely nothing went to waste.

The UK was patently in need of rebuilding, and voters wanted change. The newly elected Labour government took over from Churchill's coalition government with the promise that they would take sweeping measures to improve people's lives. The plans that Labour's leader Clement Attlee proposed were all grounded in a deep-seated devotion to social justice and represented a marked dogmatic difference from Churchill, who only months earlier had heroically announced the unconditional surrender of Nazi Germany. Labour's forward-looking election slogan, 'Let us face the future', was far more appealing than the Conservatives' plea to let Churchill 'Finish the job'. The 'Social Insurance and Allied Services' report, written by the Liberal politician William Beveridge, formed the basis of the welfare state that Attlee subsequently introduced.

Having declared that 'the Labour Party is a Socialist Party, and proud of it', Attlee outlined that there were five 'evil giants that needed tackling on the road to reconstruction'. 'Want' would be dispelled by a fair income for all. 'Disease' would be handled by granting access to healthcare for everyone. 'Ignorance' would disappear with a good education for all children. 'Squalor' would be eliminated with adequate and affordable housing for every community. Finally, 'Idleness' would be replaced with gainful employment. Labour, therefore, embarked on the creation of a comprehensive 'cradle to grave' welfare state for a populace that they believed had sacrificed so much during World War II and deserved a better future.

From 1947, Labour ambitiously rolled out their large-scale programmes, with Aneurin Bevan, the Minister for Health and Housing, leading the charge on two fronts. Inspired by the Tredegar Medical Aid Society that had been founded in the coal mining town in South Wales where he grew up, he introduced the National Health Service (NHS). The service was to provide health care that was free at the point of use in return for contributions from the taxpayer.

Bevan envisaged the social housing sector similarly to the NHS. He wanted everyone to have access to decent and affordable homes, as well as green space to cater for people's general health and recreation. Thus, in 1947, the Town and Country Planning Act was introduced, with a target of building 300,000 new houses a year. The initiative had its origins in a series of Housing Acts that had been established late in the previous century, which were originally designed to provide general housing and amenities for the working class. Labour subsequently removed explicit references in the legislation to housing for the working class and introduced the concept of 'general needs construction'. Aneurin Bevan best defined this notion with his vision that 'the working man, the doctor, and the clergyman will live in close proximity to each other'. Over a six-year period, the Homes for All policy led to over one million council houses being built for tenants to rent, while Green Belt land was designated across the country to keep specific spaces protected from development.

Committed to the vision outlined in their manifesto, Labour also introduced the nationalisation of key industries in the same year, putting everything from the railways and the airlines to the country's gas and electricity in the hands of the people, instead of a small group of shareholders. Nationalisation appealed to many workers who felt that employers, such as coal owners, had been more interested in profit than addressing the dangerous working conditions of their workers.

The new National Coal Board intended to be as much of a humanitarian institution as an economic one.[20] As part of its formation, 850 colliery owners were compensated with over £150 million, and a notice was posted at every coal mine, reading 'This colliery is now managed by the National Coal Board on behalf of the people'.[21] One such colliery was Bowes and Partners, who in 1947 handed over ownership of Monkton Coke Works to the Durham Division of the National Coal Board. Over the next few years, the National Coal Board continued to operate the plant as efficiently and effectively as possible.

By the time David had returned from war duties, Barbara was now a toddler and Jennie needed to reacquaint herself with a man who on paper was her husband, but with whom she'd had only written contact for years. This was a peculiar circumstance that housewives worldwide had to experience, and thankfully the young couple were able to reignite a bond that was to last both their lifetimes. Jennie had been concerned that her husband may never meet his first child, so it was a huge relief that he was back home safe. With David now an extra person in an already confined space that Jennie was sharing with her sister's family, the couple looked to rent a small place of their own. With little money available, they moved into an insalubrious one-bedroom flat on William Street near the quay area of Hebburn.

Jennie and David were very much contributors to the Baby Boomer generation, with birth rates in the West skyrocketing. By 1950, daughter Moira had also been born, with sons David and Brian to follow shortly thereafter. Despite cohabiting in a cramped space, the couple wanted to create a warm and loving environment for their children and they fulfilled their responsibilities remarkably well. Jennie took pride in her role as a housewife, looking after her family and keeping their home clean. David, who had first started working in the pits at the age of fourteen, was the breadwinner, earning just enough from his mining work in Wardley Colliery to provide for his family.

The Town and Country Planning Act was in full swing, with the Labour government overseeing the creation of council estates across the country. The land in the southern part of Hebburn, with the exception of Monkton Coke Works, had remained predominantly agricultural. It was one of the areas in the North East that had been chosen for residential development.

The Alkali and Clean Air Inspectorate had been an established government agency for many decades. It was responsible for implementing policies around air pollution controls with a particular focus on major industrial emitters. The inspectorate was more than aware of the importance of coke in the drive to reduce air pollution across the country. The Great Smog of London in 1952, induced by a combination of poor weather, and smoke and sulphur deposits from factories and homes using coal, had led to the deaths of thousands of people. Such a disaster could not be repeated, and the use of coke in factories and homes was a crucial way to ensure this.

For the inspector assessing the residential development plans in south Hebburn, the creation of coke was not the issue. The problem was building a housing estate next to a polluting plant and noisy machinery. On a visit to the area, the inspector warned that high-density public housing should not be constructed in such proximity to the polluting coke works.[22]

The local council, though, was on a mission to help people rebuild their post-war lives and support the government's programme. Setting the tone for generations of regimes to come, any concerns about environmental impact in the area were injudiciously disregarded by council officials, and construction began.

In 1953, Jennie and David received the marvellous news from the council that they had been selected to move from their one-bedroom flat into one of the new houses that were being built in south Hebburn, an area that was to be hitherto known as Monkton Lane Estate. The locality comprised four main streets, with Melrose Avenue converging onto Hexham Avenue and connecting to Finchale Road, with Monkton Lane and Luke's Lane running parallel. Their semi-detached property was to have three bedrooms, a garden, hot water, and an indoor toilet. This was modern luxury: a brand-new home, with space enough for three bedrooms, a patch of grass on either side of the building to grow flowers and plants, and no more outdoor netty![23]

The house that they moved into with their four children was 1 Melrose Avenue, situated at the intersection of Melrose and Hexham Avenues. On entering, the stairwell led to the upstairs bedrooms. There was a master bedroom on the right and front of the property for the parents. There was another bedroom to the left, large enough for a bunk bed and overlooking the front garden. At the back of the house was the final bedroom, laundry cupboard, and bathroom. Downstairs, the centre point of the living room was the open fireplace, with plenty of space for a sofa and two armchairs. Next, continuing through towards the end of the room was a dining area, comfortably large enough for a china cabinet and dining table. The abutting kitchen led to a plot behind the property containing an ample lawn and space for a car to be parked. Being the first building on its street, the property displayed the Melrose Avenue street sign on the exposed brick wall outside, above the living room window. In front of the main entrance was a small path connecting the door and the waist-height gate. The view from the front room, beyond their hedged garden and pathway, was that of unobstructed acres of farmland with dairy cows grazing in open space to the left, and Monkton Coke Works to the right.

> We would play with the corn stacks and throw turnips and cabbages that were growing from the ground in Hexham Avenue field opposite the house. On an evening, we'd play hide and seek among the foundations of the half-built houses and make sure we didn't get caught by the Watchie who was keeping an eye on the building sites![24]
>
> — Moira, Jennie's daughter.

While the coke ovens spoiled the bucolic landscape and raised concerns over the pollution that they were creating, the council had reassured incoming residents that the plant would shortly be closed.[25] Coke batteries have a limited lifespan due to the fierce heat and coke ash that damages their walls. The useful life of a coke battery is around 15,000 pushes. An average coking time of approximately twenty hours corresponds to a median coke battery lifetime of about twenty years. With Monkton Coke Works having begun operations in 1937, the 20-year mark was fast approaching.

The area that the Shearans had just moved into was set to become a busy residential estate, with new houses for young families being constructed all around them. Some of the homes were being built with gardens just 40 feet from the boundary fence of the coke works, the equivalent of the length of a car parking space. While incoming residents were perturbed to be living so close to the plant, they were greatly relieved that with the coke ovens coming to the end of their lifespan within a couple of years, the site would soon return to its original woodland state.

Jennie committed herself to raising her family and running the household. Each evening, once David had scrubbed off the coal dust he was covered in after a day down the pit, Jennie would have the family meal already prepared. Homemade soup using a ham shank or a stew were popular favourites with the children. David was happiest on the rare occasions that a simple smoked haddock accompanied with boiled potatoes was served. The whole family were unanimous in what they loved for Sunday dinner though; roast beef with all the trimmings, followed by rice pudding.

With one sole source of income to support an ever-growing family, Jennie conscientiously managed the budget and ensured that everyone was always well fed, clean, and well dressed for school and church. The family lived modestly. Most weekends, the family would enjoy walks to the park and visits to the coast to take in the fresh air and walk among the sand dunes. On Saturday evenings, David would wear his suit and tie and go to the nearby Mill Tavern for a couple of pints with his friends, while Jennie would have a catch-up with her neighbours in between taking care of the children.

Jennie managed the finances well, so much so that the Shearans were the first in Monkton Lane Estate to have a black and white television. It was unique to have this technology in the home, and children from the nearby houses would sit in a row next to Jennie's children to watch *The Sooty Show* and *Crackerjack!* on the one channel, the BBC.

Neighbours would often knock on the door for help, and Jennie never turned anyone away, welcoming everyone. A large part of her kindness stemmed from the philanthropic post-war socialist ideology that she identified with and championed. The Labour Party had always been about people and improving their lives, and Jennie contentedly epitomised these principles.

> She would help anybody, anyone at all. I remember one Christmas Day, we were all ready to sit down for Christmas lunch. A knock on the door came. It was a neighbour's child. 'Mrs Shearan, can you come over? Me mam's having a baby'. My mother went over and we didn't see her again until 10 o'clock at night as she was helping the doctor and the midwife.
>
> People would always be knocking on the door.
>
> 'Mrs Shearan, me mam said, can you sell her a bucket of coal?' The neighbour's child would have her

hand held out with an empty bucket. I would take it
to my mother, and, without raising her head from the
scrubbing and the polishing, she would tell me, 'fill it
up for her'. I would come round the back of the house
with the coal and take the girl's threepence, to which
my mother would order me to give it to her back. The
little girl, knowing about my mother's generosity, was
still standing there and waiting for her threepence,
when I returned to her![26]

— Moira, Jennie's daughter.

Jennie was no different to any of the mothers on Monkton Lane
Estate. The Shearans had extra coal because David worked in
the pit and would receive coal as part of his stipend. A strong
sense of community permeated the neighbourhood, with all the
households looking out for each other. World War II was not yet
a distant memory. People supported one another, using whatever
means they had available to them.

With the properties still so new, the residents took great
pride in keeping their homes pristine, and it was common for
adjacent households to help each other with their painting and
decorating. Domestic jobs were carried out by whoever in the area
was skilled in that particular task. Jennie's husband David was
well known for being dexterous at making or fixing most things
and he became the unofficial car mechanic for his neighbours.

A lifetime of manual labour had honed David's muscular
physique and made him completely at ease performing complex
tasks with his hands. He was especially talented at woodwork,

and on weekends, from his handmade eight-by-eight-foot garden shed, he would craft everything from bedside lamps to cupboards for Jennie. To drive the lathe, he used a disregarded bicycle, transforming its chain and pedal into the lathe's operator. His children and youngsters from the neighbourhood would take turns rotating the pedal as quickly as they could while David shaped his next appliance. Time in the shed at his workbench was his creative outlet, a welcome respite from the treacherous mines where he would spend his working weeks. After a fruitful morning of carpentry, he would play the mouth organ and sing amusing ditties in his Welsh brogue to his children.

> To market to market went my brother Jim,
> But somebody threw a tomato at him,
> A tomato's alright when they come in the skin,
> But this one it didn't, it came in the tin![27]

Jennie also spent a lot of time in the back garden, mostly to cultivate her many varieties of flowers and to hang the family's laundry out to dry on the washing line that ran diagonally across the lawn. While it would upset her that the freshly cleaned linens would often be spoiled by speckles of soot from Monkton Coke Works chimneys' push, she found great joy in growing her plants. Whenever she was out walking and saw a shrub that she liked, she would take a little cutting and try to grow it among her ever-growing mishmash of pots full of daisies and daffodils. David had planted rambling pink roses across the back fence, and they were especially admired by Jennie and her neighbours when in full bloom.

Life was good for Jennie. It was by no means easy, but she and David and the children were happy in their neighbourhood. On Sundays, the distinctive melodic chimes of Minchella's Italian ice cream van could be heard as it pulled up opposite their home, a stopping point for the estate's local vendors. In the early hours of the weekday mornings, the electric milk float would arrive and deliver bottles of milk quietly. Bill Johnson, who lived just a little further down the street, on Hexham Avenue, would pull up throughout the week with his mobile grocery shop, selling everything from cigarettes and Dandelion and Burdock[28] to potatoes and apples from his makeshift counter, space enough to allow only one customer in at a time.

Yet nothing elicited as much excitement among the children of the estate as Minchella's. Jennie's bairns[29] would run out to get the ice creams, and on their excited return, she would place the treats in individual glasses and pour lemonade over them to create an extra special frosted lemonade. After watching the Sunday matinee, usually a cowboy movie or a musical, Jennie would tidy up, and David would do the crossword. The children would spend the rest of the day outside until tea-time, the girls freely playing hopscotch and skipping ropes, and the boys playing Cowboys and Indians with their friends.

The field that the children would play on was directly in front of and adjacent to Monkton Coke Works. With just a boundary fence and Monkton Lane separating the children from the plant, the facility was an ever-present backdrop to their childhoods.

Towards the end of 1953, the residents received some con-cerning news. The coke batteries of Monkton Coke Works may well have been reaching the end of their natural life, but rather than dismantle them, the National Coal Board had another altogether different intention for the site. They had received a significant request for ongoing deliveries of coke to Sweden. Instead of switching off the plant, they were going to fulfil the order and build a new set of coke ovens to replace the original battery that would soon reach the end of its useful life.

In August of that year, Hebburn Urban District Council granted the Coal Board planning permission for the extension of the site in the form of an additional battery of 33 ovens to be constructed to the south-west of the original battery. The council had permitted the proposal because, in their opinion, the replacements did not materially affect the external appearance of the premises. The replacement of the original coke battery with a new set was rubber-stamped without any prior notification to residents.

This authorisation represented a significant expansion. Shortly after the new set of ovens joined the soon-to-be-defunct existing battery, planning permission was also given for the stocking of coke on the south-eastern part of the site that had formerly been an agricultural field. Monkton Coke Works was growing fast, with the arrangement of buildings becoming implacably more complicated and extensive. What was a small-scale plant when Jennie moved into 1 Melrose Avenue had very quickly more than doubled in size to become a leviathan monster that bore down on the whole community.

In 1956, The Clean Air Act was passed in response to the Great Smog of London from four years earlier. It aimed to tackle air pollution created by the burning of coal and industrial activities. Smokeless Zone regulation soon followed, banning coal fires at home in a bid to clean up the environment. Consequently, coke was heralded as a necessary smokeless fuel substitute for coal in all domestic heating, and measures were put in place to incentivise its use across the country. Hebburn would become the first town in the country to have the Smokeless Zone policy enforced, with a penalty system put in place for any non-compliant residents. The irony was not lost on Jennie.

CHAPTER THREE

DANTE'S INFERNO

'The sky was lit up with flames as they would blast out smoke.'

Monkton Coke Works had cemented itself into the landscape of Hebburn and the community of Monkton Lane Estate had no choice but to accept its presence. For the parents, the hope that the site would someday disappear was sadly fading away. For their children, who had never experienced life any differently, the plant was simply another attraction in their make-believe Monkton Lane theme park. The looming infrastructure of the facility interwove with the Bowes Railway line that emerged from the site's entrance and led to the remnants of a long since disused colliery to turn Hexham Avenue field into an adventure-filled playground free of any concerns about health or safety.

On a Saturday morning, about seven o'clock, we would all meet outside the house. I would have been around nine years old. There would be about eight of us; me, my brother David, and our mates. We used to walk from there, up the railway lines to Wardley, and up to Leam Lane. We would go to a big swimming pool there and we used to be in there for hours, and then walk home, back along the line.

The railway used to come from Wardley pit and go through Monkton and right into the coke works. Down the bottom of the field was a redundant mining works, with big wheels on the top that were used to lower the cage into the bowels of the earth for the miners. We used to play on that! You had all the big wires coming down, and on the ground level they had a long steel plate across the shaft and there was a little hole in the middle for the wire to go down into the pit. We used to drop stones in the hole; it would take ages for them to splash! We used to climb up the shaft, right the way to the top of the mining works, wriggle through the holes, then slide down the ropes. It was fabulous, but when I think of it now I get cold shivers. We used to land on the steel plate and if that had collapsed, we would have ended up in the pit. I remember one lad breaking his leg when he fell off a disused larry car.

All the while the coke ovens were going full belt. We used to play cricket and if you were rolling on the grass, you used to go home totally black off the muck from the ovens.[30]

— Brian, Jennie's son.

Evidently, the carefree children in the neighbourhood were mostly unperturbed by Monkton Coke Works and its haze of pollution during playtime. At that age, the surrounding fields were merely considered an exotic obstacle course, and the venue for many light-hearted games of football, with the children only careful not to inadvertently step in the bright yellow sulphuric ditch water that seeped from the site when their ball went out of play.

The structure of Monkton Coke Works was so dominant in the area that the older cross-country runners from the adjacent St Joseph's and St James' schools would jog around its poison-pumping perimeter for training. After they had completed the circuit, their once crisp white PE vests would be covered in black marks from the soot. Steve Cram was one of the young men who would often circulate the plant and, after leaving the area, became an Olympic medallist for middle-distance running. In later years, he would joke that the pollution from the facility inspired him to run faster.

Operating eighteen hours each day, however, the plant was thought of very differently at night by the youngsters in the area. The relentless tannoys and blazing fires would be terrifying for the children who experienced Monkton Coke Works from their bedroom window.

> On a night time you've seen nothing like it. You have to remember we only lived a few hundred yards from the boundary fence. The sky was lit up with flames as they would blast out smoke. I used to have nightmares.[31]
>
> — Brian, Jennie's son.

Jennie would be there to comfort her son from his bunk bed, reassuring him that it was just a bad dream. If only.

In 1959, Jennie was pregnant with her fifth child, unbeknownst to her younger children, and gave birth to Maria in her bedroom.

> This is how naive we were back then. Me and my brother David didn't even know my mother was having a baby, because in those days the parents didn't tell you and it just happened. We were playing football in the field and our next-door neighbour shouted on us and we ran over to the fence and he said, 'You've just got a little sister'. I looked at him and said, 'No, we've got two sisters,' and he says, 'You've got a new baby sister!' I didn't have a clue what he was on about, then we went in the house and up the stairs, and low and behold, there was Maria.[32]
>
> —Brian, Jennie's son.

Jennie was now 37 and, despite being relatively old for the time to be a new mother, was as energetic as ever. Maria, being the youngest child by some way, was adored by the family. Her parents cherished her, and Barbara, now fourteen, was especially affectionate towards her little sister.

Within a few years, Barbara was to be the first child to leave home. She had met Terry, a local Teddy Boy whose father also worked in Wardley Colliery, at the Hebburn Store Hall Saturday night dance. After a three-year courtship, they married at St

Aloysius Church. Despite Barbara's move to nearby Hedgeley Road, the Shearan household remained a buzz of activity, and a final addition to the family came in 1965 with the arrival of Sally, a gentle and intelligent German Shepherd with a smooth black and tan coat. Despite her sizeable frame, Sally was sweet-tempered and docile, and children from neighbouring houses would often ask if they could play with her.

> This one time they took Sally onto Hexham Avenue field to throw the stick with her. One of the kids accidentally threw the stick too far and it ended up in a slurry pit. The dog went straight in it and got into great difficulty. It was like quicksand. Fortunately, there was a man at the coke works who was seeing what was going on and he had a rope and a noose on the end of it and managed to pull Sally out. She just legged it home and ran across Hexham Avenue field, straight through the house and up to the back garden. She splashed everywhere, dirtying the sofa and wallpaper. The dog was completely black. My father initially chased her out because he didn't even recognise it was Sally. He had to tie her up and hose her down. The poor dog was being sick and bringing all the black crap up.[33]

> — Brian, Jennie's son.

This is what life had become on Monkton Lane Estate. The effluent that had tarnished and terrified Sally would eventually reach everyone in some way.

In 1964, much to the confusion of the neighbourhood, Hebburn Urban District Council recklessly gave permission for another large new housing estate to be constructed near the coke works. Running directly opposite the plant, Luke's Lane Estate was built against the strong advice of the Alkali and Clean Air Inspectorate.[34] Once again, the drive to create housing overrode concerns for the well-being of the inhabitants. It appeared that the environment was not so much an afterthought, but that it was not being given any thought at all. With the number of residents living close to the site about to rise dramatically, there was no alternative but to acknowledge that Monkton Coke Works would continue to play an oppressive role in people's lives in the ever-growing area.

Employing a fluctuating figure of around 200 workers across the 1960s, the plant's output was outstanding, producing 285 thousand tonnes of carbonised coal and stocking over 250,000 tonnes of coke each year.[35] From the perspective of the National Coal Board, whose self-proclaimed modus operandi was to manage the coal industry in a humanitarian manner, Monkton Coke Works was a successful operation. It was clear though that any semblance of humanitarianism only extended to within the walls of the plant, and no further. The older original battery of ovens, which had finally reached the end of its life in 1960, had still yet to be dismantled. It served as both a brutal and Brutalist visual reminder of the site's intransigence for those living on its doorstep.

In the 1960s, the National Coal Board filmed a short documentary to depict working life within the plant.[36] Opening with a frame stating who its owners are, the grainy, flickering silent footage proceeds with a long shot showing a section of the site. With just a handful of the blast furnaces, chimneys, and ovens on display, some unassuming wisps of smoke coming out of the smallest chimney dissipate completely before they get a chance to reach the gentle cirrocumulus clouds in the pale blue sky. The boundary fence and agricultural field in the foreground suggest to the viewer that the site is operating entirely independently, miles away from any residential buildings. The next shot cuts to a well-kept flower bed, with flourishing tulips gently swaying in the wind as the camera pans to some outbuildings. The viewer is then introduced to a series of frames that establishes the enormity of the site, from huge barrels to an unceasing procession of seven-foot larry cars carrying mounds of coal. A lone man with a flat cap and a metal rod in his hand walks in between the two rails of coal trucks and greets a colleague as the trucks enter a shed, providing the viewer with a scale of how big the factory is.

A new section is presented: 'Charging coal into ovens'. A large moving platform with four considerable metal containers on top of it moves along a set of rails. It pauses as an employee in dungarees watches five chutes underneath the platform discharge coal into openings in the floor of the gantry. Heavy black smoke billows from the ground, while bursts of flames indiscriminately appear along the rails. The large undulating mass of fumes turns dark grey, as the earnest labourer, face fully exposed to the discharge, assuredly and expeditiously opens a vent on top of a cylindrical oven to let out yet more flames and heavy smoke.

Another man, eyes, nose, and ears also unprotected, performs the same task concurrently. The view is next shown from the top of the cylindrical ovens and a workman opens a vent on top of one of them. Flames continually shoot out, as smoke blows out of the platform floor. The rugged men with craggy skin are at total ease in their surroundings, diligently operating the devices around them.

The momentum of the documentary gathers pace as it switches to more and more individuals performing their specific tasks. One operator uses a hand-cranked device to lock a large steel block in place on a coke oven door. When the hatch is opened it releases a smattering of coke pieces. The faded Technicolour of the frames bursts into life as the coke glows with the heat.

A second title is shown on the screen: 'Pushing coke from the ovens'. A close-up follows. This time, a worker on the mobile gantry operates the ram to push the coke into the oven. The intense heat of the red-hot furnace is revealed by the glare of the ram as it is withdrawn. A never-ending avalanche of new glowing coke cascades into a large container. It is a breathtaking sight. As the coke hits the receptacle, clouds of smoke emanate from its base. The coke is taken to be quenched with water, as the lens zooms into the incandescent orange-yellow hue of the coruscating embers. The coke arrives at the shed and the camera tilts upwards to display a towering brick flue, with steam and smoke pouring out of the chimney, taking over the screen. Following the journey of the newly created coke breeze, a row of grill-like structures is shown. The closely spaced dividers hold back the recently quenched coke as a workman controls their descent down a chute. A conveyor belt then takes the coke to be loaded onto rail trucks for industrial and domestic use. The documentary ends with a shot of the facility once again surrounded by the green field.

It is a powerful piece of well-edited propaganda for Monkton Coke Works. The omission of sound, probably due to technological limitations at the time, fails to indicate to the audience the harsh noise that the plant was generating. The extensive grassland and blooming flowers suggest vast acres of terrain between Monkton Coke Works and the general populace which, of course, was not the case. Combined with the peaceful opening images, inoffensive quantities of smoke in the following frames disingenuously encourage the observer to consider this an inoffensive plant that does not negatively impact the surrounding environment. The scenes of the employees working in harmony with the machinery propose a quixotic version of a well-managed coking process that focuses only on the engineering might of the plant, entirely disregarding its proximity to the local community. It is a calculated advertisement to the proletariat about the might of the National Coal Board, and a befitting emblem of the potency of this institution in the 1960s.

Indeed, for employees inside the plant, the lack of personal protective equipment that the documentary reveals made for an unpleasant working environment.

> The conveyor belt for the coal came down and it was gradually fed to the oven. My job was to start at the top, the seventh storey, and sweep all the coal dust from one floor to the next floor, to the next floor, to the next floor. By the time we'd reached the bottom,

we were absolutely covered in coal dust, and it was ready to go again. No safety clothes were provided. There were no face masks, no eye shields, no high-vis jackets. Nothing at all. When I got home I used to have to get washed twice as I was so dirty.[37]

−Billy Corr, former worker at Monkton Coke Works.

Time was moving on. The neighbourhood was getting older. Jennie had become a grandmother for the first time, with Barbara having given birth to a daughter. Jennie was taking care of her husband David, who had suffered a serious accident in the mine from a loose heavy stone paralysing his left arm. Moira, very much embracing swinging sixties fashion, was now working in a women's clothing store, and engaged to marry her partner Ian. Son David had started his plumbing apprenticeship, and Brian and Maria were studying at secondary school. Sally was slowing down. Beatlemania came and went. The Conservatives swapped power with Labour, and back again. Black and white televisions were replaced with colour sets and there were now three channels.

The only constant was the Cokeys.[38]

Jennie was feeling a growing sense of frustration. Like all the other incoming young families in 1953, she had moved into Monkton Lane Estate, elated and grateful for her new idyllic abode. At the time, many residents had expressed their concerns about the plant, with Jennie's home being only a few hundred yards from the perimeter fence of the site.[39] Their minds had been put at rest with the council's reassurance that Monkton

Coke Works would soon cease operations. Nearly twenty years had gone by, and their utopia was now a dystopia; the monstrous ovens outside their homes showing no signs of abating, while the council's pledge had revealed itself to be misleading.

By the time Jennie was 50, all the Shearan children, with the exception of thirteen-year-old Maria, had left home to start families of their own. Jennie devoted herself to raising her daughter and looking after her husband. The casualty that David had endured at Wardley Colliery had dislocated both of his shoulders, tore two inches of brachial plexus nerves in his left arm, and crushed his left hand. Such was the physical damage that was incurred, his visits to the Durham Miners' Rehabilitation Centre lasted over two years. Despite the Centre warning that David would never again be able to move his hand, which had shaped into a claw, Jennie assiduously massaged his paralysed arm each night, and he eventually regained feeling in his fingers. Sadly though, David's beloved maroon Ford Consul had to go, as did the freedom that a car offers. On the daily walk that Jennie would make to the bus stop, to run errands and go shopping, Monkton Coke Works was ubiquitous. What had started as a limited-life facility, was now a deep-rooted polluter that was here to stay.

Jennie could not avoid seeing the grotesque geometry of the monochrome beams and ovens as she walked down her path. On her way to catch the Finchale Road bus, she would temporarily forget about the bullying chimneys and bump into a neighbour. 'Hello Maudy, how are you keeping?' Their conversation would be interrupted by the acrid stench of the plant's sulphuric emissions. Jennie would arrive at the bustling streets of the New Town and browse through the shopping centre, particularly enjoying a visit to the furniture store Callers or getting her hair styled at the hairdresser. Jennie's last stopping place would be Walter Willson's supermarket. Walter Willson had opened an eponymous chain of grocery stores in each pit village in the region and was an established name in the New Town. For a small fee, 'The Smiling Service Store'[40] delivered the items that the customer had purchased, often just as they had returned to their house. As Jennie approached Monkton Lane Estate, always sitting at the bottom level of the double-decker, the heavy dark clouds of smog that the facility produced, enveloping the rooftops that stood only a few yards from its boundary fence, were the unmistakable signal that she was approaching home.[41]

Monkton Coke Works had infiltrated every part of her and her neighbours' lives, and there was no way to escape. It seemed from the outside that the plant was a prolific and unstoppable behemoth, focused purely on fulfilling its sole purpose, regardless of any external considerations.

Jennie's vexation towards an institution that appeared to operate wholly on its own terms was now brewing into anger. Looking for a way to resolve the predicament that she and her town were in, she turned to local politics.

The constituency of Jarrow is a Labour stronghold that comprises Jarrow and Hebburn. When the town's politicians would visit the 3,000 residents of Monkton Lane Estate canvassing for votes, Jennie would debate ardently with them. She wanted to understand the political machine, and what could be done to help the people in her neighbourhood. She started regularly attending the local Women's Labour Party meetings on Wednesday evenings and was such a valuable and vocal participant in the sessions that her fellow attendees suggested that she compete in the local council elections.

After his mining accident, David had only been given a low monthly lifetime disability pension, and in 1972, when the National Coal Board made him redundant, the severance package and pension he received was also insubstantial. Aged 53, David wanted to keep working, and secured a job at a ship-builder, where labour conditions were far less onerous than at the colliery. For Jennie, this felt like a moment to fulfil an ambition that she had held since her teenage years. This was her time to get involved in politics, to 'get into a position where you could help your fellow man'.[42]

In 1973, aged 51, Jennie put herself forward as Labour Party County Councillor for Hebburn. Management for the district of Hebburn was imminently going to be transferred from Durham County Council to a newly formed county, called the Borough of South Tyneside. This would now comprise the boroughs of South Shields and Jarrow along with the smaller urban districts of Hebburn and Boldon. South Tyneside Council would now be the Local Authority that provided most of local government services to the area, running departments ranging from Planning and Education, to Transport and Environmental Health.

The Shields Gazette, the country's oldest provincial evening newspaper, reported daily on all matters regarding South Tyne-side, and was especially prolific regarding matters involving local politics. A faded clipping from the inside of the *Gazette*, as it was more commonly known, shows a row of candidates in the forthcoming Tyne and Wear Metropolitan County Council elections. Among a set of three images of suited men fighting for the Jarrow and Boldon seats is a smiling Jennie with bouffant hair, head tilted, per the style of portrait photos of the era, wearing a polo neck jumper. Described as 'newcomer to the political scene, Mrs Jennie Shearan, a housewife', this was Jennie's first foray into the world of politics and media. Alongside her for the campaign was colleague Dick Fenwick, who had served on Hebburn Urban District Council for over two decades. The newspaper accurately predicted that their campaign would be a 'walk-over',[43] and Jennie was duly elected by the town to become Hebburn's representative on the county council. Her responsibility covered the whole town, from the quay area where she grew up, to the New Town district where she shopped, and the southern part of Hebburn where she lived. Once the results were in, a different article proclaimed, 'Jennie takes a seat with the experts'. She was now one of only 12 women among over 100 men serving as a councillor in the county, and the only woman without any previous local government experience.

Coming into the role, Jennie immediately identified her first issue. General awareness within the town was low about who the councillors even were.

> Because I have never been a councillor before, I keep wondering whether people know me. On Thursday, when the Hebburn elections took place, people kept asking me who the councillors were. I felt very embarrassed telling them I was one.[44]
>
> —Jennie Shearan.

Jennie wanted her role to have relevance within Hebburn, and to have a positive impact on peoples' lives. As a democratically elected community representative, she understood that she had to be visible within the town and be regarded as a trusted point of contact. 'Battling Jennie'[45] threw herself into becoming the spokesperson for the town she loved, speaking and working with members of the public and community groups to understand their needs. She immediately began making regular appearances in the local newspapers with small features recognising her widespread efforts to improve her community.[46] She worked with the Tyne and Wear Transport Committee to install a pedestrian ramp at Hebburn Railway Station, chiefly with the disabled and young mothers pushing prams in mind.[47]

'Tyne and Wear consumer watchdog, Mrs Jennie Shearan'[48] launched the Tyneside Consumers' Union to address inflation concerns and customer service issues. She began monitoring the prices of commodities in the town's stores and overseeing refunds for faulty merchandise sold to local shoppers.[49] Jennie even challenged a television manufacturer for a fair deal for tenants in the town.[50] This was a woman of principle who was not afraid to go up against established companies for causes that she believed in.

> Being a councillor is one of the greatest honours anyone can bestow on you. People make you a councillor because they have faith in you and you haven't to let that faith down. You are the public's servant.[51]
>
> — Jennie Shearan.

She was doing work that she loved, she was good at it, and she was fighting for causes that gave her constituents a better quality of life. In becoming an effective advocate for the best interests of her ward, Jennie was learning so much. Despite leaving school at an early age and receiving only a basic education, she was very bright, and it showed in the impact that she was having on Hebburn. The quotes from the newspaper clippings suggest a woman who was becoming well versed in leveraging the local media to raise attention to issues affecting her community. She was also demonstrating an understanding of how to run campaigns and how to influence key decision makers. Jennie was navigating the complex landscape of local public services and building a valuable network of contacts throughout Tyneside. Perhaps most importantly, her eloquence and colloquialisms revealed her kind disposition and authenticity. Her consistent results and trenchant communication style were winning many admirers.

Jennie's achievements were all delivered at a time when men had an inordinate share of voice in politics. As soon as Jennie became councillor, she realised that she had entered a boy's club, and was eager to add more women voices to the sole dozen on the Tyne and Wear Metropolitan Authority. 'I would like to add another. Thirteen is my lucky number. But thirteen or eleven, all the women are going to find they have a lot to do.'[52] She encountered this gender inequality throughout her time on the council. One article reported Jennie, now 55, becoming a new member of the North Eastern Sea Fisheries Committee, the first woman to be elected to the body.

I felt a bit awkward when I came in and saw all those
men but I soon settled in and I found the whole work
interesting and important. I'm pleased really. I think
there should be far more women on all public bodies,
from Parliament down.[53]

— Jennie Shearan.

One particular issue that secured a lot of coverage in the local press
underlines the reputation that Jennie had by now established. The
lead photo in one article shows her striding towards the camera,
absorbed by the task in hand. She is accompanied by a crowd
of mostly male local councillors, development delegates, and
town planners. The article begins, 'When Hebburn's County
Councillor Jennie Shearan investigates a complaint, she really
does get down to business'.[54] The quay area of Hebburn had been
blighted with rubble tips left on the streets, and derelict land in
need of landscaping. Jennie saw the unfairness in this for the
residents and, as one journalist wrote, 'launched a one-woman
campaign'[55] to tidy up the area.

I was born and bred in this area and I think it has
been neglected. I had to get down on my hands and
knees to find the drains in the street … Quite honestly,
the area is disgusting. New council houses have been
built, but if these had been private houses this mess
would never have been allowed … I will not leave it
alone until I get some money for South Tyneside. The

people simply want the street corners cleaned up and the grass verges made nice.[56]

— Jennie Shearan.

Prime Minister Clement Attlee and Minister for Health and Housing Aneurin Bevan had long since left their leadership roles in the British Government, but their determination to eliminate 'Squalor' was ingrained into Jennie. Her persistent lobbying worked. The *Gazette* labelled Jennie 'Cinderella'[57] when Tyne and Wear's Environment Improvement Committee agreed to allocate funding of £10,000 to the clean-up effort.

While the seriousness of politics was increasingly becoming the focus in Jennie's life, there were still plenty of moments of levity when the community got together. In June 1977, the country celebrated the Queen's Silver Jubilee, an occasion to mark the 25th anniversary of Queen Elizabeth II's accession to the throne. Street parties were organised throughout the UK, and Hebburn was no exception. Bunting was strung throughout Melrose and Hexham Avenues, and a long row of tables filled with scones, cakes, and trifles were assembled at the junction for the whole estate to commemorate the event. After the egg and spoon race and the sack race, the estate's children participated in a fancy dress competition. In third place was Miss Mopp, an archetypal house cleaner, with rollers in her hair and a headscarf. The winner was a very authentic-looking suffragette, but it was second-placed Glenn, Jennie's five-year-old grandson, who featured in the local paper the next day, dressed as a coal miner, topless but for a chequered scarf and black shorts, soot on his face, and a miner's lamp in his hand. The photo and headline in the *Gazette* was of a smiling Glenn, with the National Coal Board branded larry cars of Monkton Coke Works in the background, below the caption reading, 'Me granda was a pit yacker'.[58]

Throughout her life, Jennie had always loved children and she saw it as her responsibility as Hebburn's County Councillor to guide them towards a better future. She had started to notice some delinquency in the town and with typical empathy sought to both understand and address the anti-social behaviour. One Bank Holiday Monday, Jennie organised a party for all the children of Monkton Lane Estate and invited the press along so that the story could inspire other nearby estates to do the same. She brought in the help of South Tyneside's Cultural and Leisure Activities Department to provide the equipment and, with the help of the community's parents, they set up a tea party in Hexham Avenue field. There was an assortment of games and races for the children, a knobbly knees contest for the parents, and a disco to round off the afternoon. Her interview with the local newspaper summarised the value that such an event brought to the neighbourhood.

> If you can provide something for the young people to do, especially during school holidays, then there will be no vandalism. Young people only turn to vandalism when they are bored, but if they have something to look forward to and to work for there is no problem. It was a wonderful afternoon and the spirit was really very good. I would recommend anyone else to have a go and discover the satisfaction they will get.[59]
>
> — Jennie Shearan.

Running parallel to Jennie's blossoming political career, was the development of another woman's passage through Parliament at the opposite end of the country. By 1975, Margaret Thatcher had become the first female leader of a major party in the UK. Her political philosophy was diametrically opposed to Jennie's.

Jennie epitomised the inclusive and fostered a community spirit for the common good. She loved the people of her town and actively wanted to improve their lives, especially the younger generations. This was antithetical to Thatcher, who felt there was 'no such thing as society',[60] and who held an instinctive hostility towards collective action. She pursued the exclusive, promoting the competitive self-enrichment of the individual.

Getting the bus for Jennie was a peaceful part of her day, and a chance for her to bump into her friends around the neighbourhood. Thatcher considered you a failure if you were using public transport beyond your mid-twenties.[61]

Where Jennie cared about advancing women in politics, Thatcher had no sympathy for 'women of ambition'[62] and surrounded her Cabinet with men from privileged backgrounds.[63]

Both women were captivating orators, speaking in an equally intelligent, measured, and cogent manner. However, Jennie's lexicon exhibited a high level of compassion for the causes that mattered to her, with Thatcher's speech frequently rooted in cool logic.

While Jennie was trying to understand how Monkton Coke Works was able to operate with so little regard for the human beings living in its vicinity, Thatcher intended to propagate deregulation. Jennie had witnessed first hand the danger of working in the mines and empathised with the workers' requests for higher pay. Meanwhile, Thatcher dispassionately aspired to reduce the power and influence of the trade unions.

The people of Monkton Lane Estate had all benefitted from Labour's comprehensive nationalisation programme. The home that Jennie lived in, the schooling that her children received, and the NHS that took care of her family's health needs, were all thanks to provisions from the welfare state. Her guiding principles were synonymous with socialism. Conversely, Thatcher held a firm belief in free-market capitalism. She sought to undo the policymaking of Attlee and Bevan and drive independence of the individual from the state. She was to become 'Thatcher the Milk Snatcher' who, confronted with a national economic crisis, worked to reduce the state's influence in the economy and privatise state-owned companies. This moniker would summarise a school of thought that would progressively starve the public sector bodies of resources. Her neoliberalist mindset focused primarily on profit, and thus considered the environment to be expendable. Thatcher's paradigm of privatisation and individual wealth creation was completely at odds with Jennie's worldview whereby the public good should be prioritised above exclusive financial gain.

Thatcher was not yet running the country, but she had made her goals clear. Labour was facing considerable challenges. High interest rates, high unemployment, and high inflation intertwined to produce that most undesirable of economic phenomena: stagflation. What sealed the fate of the Labour Party, and laid the road for Thatcher's Conservatives to take over, was the Winter of Discontent. From November 1978 to February 1979, there were widespread strikes by both private and public sector trade unions demanding pay rises greater than the limits that the

Labour government had been imposing. Lorry drivers, waste collectors, hospital workers, and even grave diggers all went on strike. Chaos was mounting, and the public inconvenience was exacerbated by the coldest winter for sixteen years, with severe storms isolating many remote areas of the country. When a general election was called a few months later, the Conservatives won a resounding victory.

In coming to power, Margaret Thatcher inherited responsibility for all the ministerial departments in the country. Among the administrations under her purview were Her Majesty's Inspectorate of Pollution and the National Coal Board. While Margaret Thatcher was now ultimately responsible for their actions, they had long-standing deliverables in place well before she came to power.

The Inspectorate of Pollution was the successor to the Alkali and Clean Air Inspectorate, and its headquarters were based in Westminster in Central London. As part of the Department of the Environment, the inspectorate's sole purpose was to regulate the most harmful polluting substances affecting air, water, and land in England and Wales. A key part of this remit was to undertake regular inspections of the most polluted areas of the country and ensure proper systems and procedures were in place to reduce emissions. The headquarters for the National Coal Board, created decades earlier by Clement Attlee's postwar Labour government, overlooked the River Thames in central London. One of its subsidiaries was National Smokeless Fuels, which focused on the coke sector.

In the same year that Margaret Thatcher was elected Prime Minister, National Smokeless Fuels finally demolished the original ovens at Monkton Coke Works that had been out of use for decades. They had just been fined £2,500 for a two-mile slick that had escaped from the plant and drifted into the River Tyne. National Smokeless Fuels had a team of well-paid lawyers who were adept at minimising PR issues and maximising legal loopholes. In a deft planning ploy, they built an additional 33 coke ovens to replace the original set. Unwittingly, Hebburn Council's planning department had stated that no planning permission was required to bulldoze the redundant battery and construct an additional batch.[64]

Thus in 1980, National Smokeless Fuels had managed to gain planning permission to double the amount of functioning battery ovens from its original number of 33 to 66.[65] An additional by-product plant was quickly constructed to manage the extra quantities of gas and chemicals that were about to be produced. The introduction of the new battery of ovens would more than double the amount of coal that was able to be processed. The annual coal throughput from the plant now totalled over 500,000 tonnes. Once again, the Local Planning Authority had supported the Coal Board chiefs, with no public participation on the decision deemed necessary.

Spurred by the council's compliance, the National Coal Board moved to gain permission for a joint venture to build a new plant producing glass bottles next to Monkton Coke Works. The news was met with hostility in the local area, with one local councillor stating, 'it seems that if a blade of grass is found in Hebburn, they will put a factory on it'.[66] Regardless, the need for employment overrode concerns for nearby residents, and planning consent was granted. The project was ultimately shelved due to a combination of reduced domestic demand for glass bottles, insufficient loan funding, and intensifying competition from abroad.

Nevertheless, despite years of promises to the contrary, once again Monkton Coke Works not only continued operations but grew significantly in size.[67] The production of domestic, industrial, and foundry coke rose from 200,000 to 600,000 tonnes a year. The tripling of production from the considerable expansion programme led to the tripling of pollution. With no legal limits to the amount of atmospheric pollution that a coke works could produce, the amount of airborne fumes and dirt dramatically increased.

Within National Smokeless Fuel's planning application for the new battery ovens, there was a written commitment to install adequate grit arrestor panels. This equipment was designed to reduce the amount of dust particles being carried into the nearby community by steam, by collecting them before they were released into the atmosphere. However, National Smokeless Fuels did not install them because the cost of the equipment would have eroded profits.

Jennie was by now conversant in the language of coking and understood that a new battery of ovens meant at least another twenty years of hazardous pollution for Monkton Lane Estate. The people of Monkton Lane Estate had been abandoned. Jennie may well have been sick and tired of Monkton Coke Works, but she was starting to see people around her *actually get sick and tired*, and it was frightening. Her dear David was prematurely weakened after a lifetime of inhaling coal dust and enduring punishing conditions down the pit. He was suffering with emphysema, a lung condition that causes shortness of breath, bursitis in his knee that left his joints inflamed, coronary thrombosis, and occupational deafness. At age 60, he suffered what was to be the first of several strokes and heart attacks.

Jennie was also beginning to notice many of her neighbours struggling with their health. She was convinced that Monkton Coke Works were exacerbating, if not contributing to, their conditions. Jennie had been fourteen years old at the time of the Jarrow March, old enough to remember the hardship that the marchers had experienced to get something like Monkton Coke Works constructed. But what Jennie was witnessing now was something even bigger than a fight for employment. She was seeing people fight for their lives. This was not about work. This was about people's health.

When Jennie was a young girl, she had been inspired by the charismatic Ellen Wilkinson and the suffragettes who fought for the right for women to vote. Jennie's role as Hebburn's councillor had enlightened her on how to navigate hierarchies and diffuse a powerful message to a broad number of people. The leader of her country was a woman, for the first time in the country's history. Jennie had accumulated a lifetime of experience of women refusing to take no for an answer in a man's world. Monkton Coke Works had become a permanent reminder that she and her neighbours were being completely ignored. Now she needed to give them a voice. This was her calling.

By now Jennie was 57 years old, and her husband needed a substantial amount of her care. But enough was enough. For over twenty years, her family and neighbours had lived with the omnipresent hazardous dirt that the plant produced, soot covering everything they owned. From her living room window, she could see the larry cars delivering the pulverised coal to the ovens. She breathed in the sulphur and felt the acid rain that the ovens produced. She was woken up nightly by the hooters, sirens, and wagons. Something had to be done.

With the local council unwilling to prioritise her community, the quality of life of her people had been degraded by institutions far removed from Monkton Lane Estate, who considered the residents out of sight, out of mind. In the decades since Monkton Coke Works was constructed, it had mutated into an ever-expanding polluting monster. Jennie realised that she had no choice but to take the matter into her own hands and fight for her community's right to clean air. With six years of proficiently fulfilling her duties as Hebburn's councillor, Jennie was in a unique position to represent Monkton Lane Estate and fully leverage all the skill and experience she had accumulated. She was not daunted by this. She was going to take them on.

Jennie with her father

Jennie and David in their younger years

Jennie with baby Barbara

Jennie with her family

Jennie's view of Monkton Coke Works from her home

POLLING DAY:

**THURSDAY
MAY 5th, 1977**

Polling Hours: 8 a.m. to 9 p.m.

You will be notified about your polling number by post

SHEARAN **X**

PLEASE DISPLAY IN YOUR WINDOW. THANK YOU!

SHEARAN

LABOUR

Jennie becomes a Labour councillor

PART TWO

PEOPLE POWER

CHAPTER FOUR

RESIDENTS UNITE

'You can move mountains with people power.'

'Tyneside shoppers' champion'[68] Jennie was by now an established face on the regional political scene, with headlines in the local newspapers continuing to celebrate her efforts to secure better pricing for goods in her town, among many other initiatives. In the spring of 1980, as she was getting ready for another day that would involve making sure David was looked after and council work was taken care of, she noticed a small lump on her left breast.

On visiting her doctor, he confirmed Jennie's worst fears. It was cancer. They had caught it early, but it needed to be taken out, now. Such were her priorities at the time, Jennie, who was in the middle of campaigning to be re-elected on the council, requested that the operation take place in three weeks' time after completing her canvassing. The doctor insisted that she urgently have the mastectomy, so Jennie negotiated a compromise. She was to be operated on at South Tyneside District Hospital once she had completed the majority of her campaigning. Bringing with her a figurine of St Anthony, to whom she regularly prayed, Jennie went to the hospital for the surgery. The operation was a success, with the local newspaper proclaiming, 'Jenny fights on!':

> She has just been allowed back to her home in Melrose Avenue, Hebburn, and in typical style promises to be back in the thick of council work in the New Year.
>
> 'I have been overwhelmed by the number of people ringing the hospital to see how I was,' Jenny said. 'In fact, I was rather ashamed because it was putting extra work on the hard-pressed hospital staff. However, I would like to thank all those who took the trouble to inquire about my health. And I will never be able to thank enough the staff and everyone at the hospital for being so marvellous. I think you have to be in hospital to really appreciate the job they do.'[69]

Jennie was sacrificing a lot; time for herself, time with her family, and her own health. At the forefront of her mind were the people of Hebburn. She was completely focused on leading and uplifting her community, and was increasingly fixated on confronting her biggest challenge yet: Monkton Coke Works.

Jennie started to collect evidence of what she saw were the detrimental effects of the plant. Unable to determine a way to collect air samples, she plucked her garden trowel and headed towards the fields next to the plant. She began collecting samples of soil and waste sludge that had escaped into the drains and run into the streets. Just as she had got on her hands and knees to help clean up the quay area, she was once again going beyond the call of duty to rid her town of 'Squalor'.

Alongside the soil samples, Jennie would collect all the fine black dust that coated the sideboards and windowsills of her house. She started taking a copious number of photos, and saved up for a camcorder to record the coke works when they were at their worst. Jennie also started to study the history of the senseless decision-making that had led to this predicament for her neighbourhood.[70]

After extensive research into archived newspapers and council records in South Tyneside Library, she found an agenda from a 1963 County Planning Committee. Shockingly, within the minutes of the meeting, there were strong recommendations not to allow new homes to be built in the area. Experts had warned that the Luke's Lane site would be subject to the full effect of pollution from the plant for about sixty per cent of the year due to adverse westerly winds in the area.[71] The National Coal Board also expressed serious doubts about the site's suitability for residential development, stating, 'we do not feel that it is desirable to erect houses in close proximity to a coking plant'.[72]

> I don't want to give the impression that I am against the coke works itself, that were here before the houses were built and at a time like this we need all the jobs we can get. But I have letters written by Her Majesty's District Alkali and Clean Air Inspector, Dr Trevor Hugh to the old Durham County Council in

1963 telling them not to build houses within half a mile of the works. I also have a letter from the Coal Board telling the council not to build close to it. Yet the council said it was short of land and ignored that advice and since then the issue had been clouded by the reorganisation of local government so no one is sure whose responsibility it is.[73]

— Jennie Shearan.

While unclear on where the culpability lay, Jennie was certain that the residents of Monkton Lane Estate were the ultimate victims. Her belief was that the people living on Melrose Avenue, Hexham Avenue, and Finchale Road should receive substantial rent and rate rebates as compensation for the air and noise pollution from the nearby Monkton Coke Works that blighted their lives on a daily basis.[74]

Jennie had by now become adept at manoeuvring through the meandering tracts of the body politic, and organised for District Value Services to visit the estate. This agency provided independent and impartial advice across the entire public sector in the UK. On witnessing the proximity of the housing to the plant, District Value Services immediately understood the residents' turmoil.

An inquiry was arranged to determine how the council could help the residents.[75] The inquiry was set to take place at South Shields Town Hall in June 1981. It was a well-known venue in South Tyneside, catching the eye of passers-by with a statue of Queen Victoria outside its entrance and a set of imposing steps leading to the main door. With its distinct clock tower and a lavish number of carvings and flowers surrounding an ornate fountain, the formality of this Grade II listed building was an appropriate setting in which to discuss a matter of such importance. In attendance alongside Jennie was her daughter Barbara, and several close friends from Monkton Lane Estate. After hours of debate, the rebates for the residents were green-lit. As Jennie celebrated the decision with her neighbours outside the town hall, journalists were there to capture the moment, and the next morning the news was in the local paper.[76]

With seven years of representing her town as County Councillor, Jennie was now putting into practice all her experience in navigating the sinuous hierarchies within local government, and getting her voice heard in the local newspaper. She knew where to go, who to talk to, how to get things done, and how to get it publicised. Jennie was thinking creatively and ambitiously, and would not settle for merely satisfactory results when it came to getting environmental justice for her community.[77]

Jennie started attending European constituency Labour meetings to learn if there was any support that could be gained at a regional level. It was at one such meeting that she met Joyce Quin and Ken Collins. Quin was the Member of European Parliament representing South Tyne and Wear and Labour, and Collins was the Chairman of the Environment Committee of the European Parliament. Ken was a recognised authority on European environment policy and much of European Union environment legislation was strongly influenced by him and his committee.

It is a start in the right direction. In July I will be leading the fight at a tribunal to get the level of [rates] reduction raised still further.

We want a further reduction, but what we would really like to see at the end of the day is a cleaner environment. This is not doing our health any good.

The coke works are now pushing ovens seven days a week, so it is obviously much worse than before. It is alright for people sitting behind desks, saying that it is no worse, but the people around here know differently. There are people who are suffering because of the dust in their beds and in their food. As well as the pollution, there is noise from Monkton Coke Works and from lorries at all hours of the day and night.

I met with the European Members of Parliament Joyce Quin and Ken Collins, and asked them if there was anything they could do. I think that there is an Environment Committee in the European Economic Community that pays compensation to people who live near factories that have a lot of pollution.[78]

— Jennie Shearan.

The early 1980s was a time when many communities needed financial support. The rebates that Jennie had negotiated really mattered for the residents of Monkton Lane Estate and were delivered while the government were raising taxes and trying to reduce the country's budget deficit. Thatcher had inherited an economy with inflation in double figures and by the early part of the decade, the UK had entered a recession.

Unemployment rose to three million and interest rates remained well above ten per cent.[79] Thatcher was under pressure to change the country's course. She had been heavily influenced by the Monetarism concept that in order to control and reduce inflation, government had to control the money supply, and to do so, any government deficit had to be reduced. Extreme deflationary policies were implemented. Although the fiscal measures succeeded in reducing inflation, it was at the cost of falling demand for British goods and lower economic growth. Regardless, the government continued to pursue its deflationary policies in a bid to achieve prosperity. Thatcher's inflexibility on the matter was encapsulated by her truculent statement, 'You turn if you want to, but the lady's not for turning'.[80] She knew she was polarising, and was not afraid to stand her ground.

Concurrently, Thatcher's resolve to push through economic transformation saw her dismantle the country's traditional heavy industries, none more so than the coal mining industry. With most of the best coal seams having been mined out decades earlier, Thatcher deemed the sector 'uneconomic' and not part of the country's revival plans. She was not alone in this sentiment. Her Labour predecessors had overseen the closure of a vast number of state-owned pits across the previous decade.

However, it was Thatcher's subsequent policies that were most disruptive for the country. She saw the state-owned coal industry as too heavily reliant on government subsidies, and its trade unions, who she dubbed as 'the enemy within', as too influential. In March 1984, the National Coal Board announced its plan to cut the nation's coal output by four million tonnes and oversee the closure of 'uneconomic' collieries. Arthur Scargill, the president of the National Union of Mineworkers, estimated that this would result in the loss of 20,000 jobs and immediately called a nationwide strike.

Controversially, not all the country's 170,000 miners were on board with the walkout, and some kept working. Pickets were organised across the country, with protesters congregating in a line outside the collieries to dissuade others from entering. There was enormous tension between picketing workers and members of the union who opposed the walkout. They were disparagingly branded 'scabs' by the picketers. It was a sad indication of an emerging fractionalisation of a once cohesive community.

Scargill had successfully led pickets during the 1970s, and Thatcher had witnessed the havoc that it had wreaked on her predecessors. This time though, she was prepared, and ahead of the National Coal Board's announcement, had taken steps to stockpile enough coal and coke to keep the country supplied for at least six months in case of a strike. She had also struck deals with non-unionised drivers to transport the coal, ensuring that power outages would not cripple the country as during previous strikes. The 1984 strikes continued into the following year, with miners violently clashing frequently with police, who had Thatcher's full support.

The widespread social unrest was rarely out of the front pages of the national newspapers. While leaders from both sides fiercely debated their opinions on the matter, communities throughout the country pulled together to support the families most affected by the strike. With no money coming into households across the country for over 300 days, funds were desperately needed to overcome the extreme hardship from which many were now suffering. The men at Monkton Coke Works were among those on strike, and a Wives Support Group was formed. For all the pollution and disruption that the plant created, the nearby community could distinguish between the company that was running the plant, and the men who worked there. In a typical display of integrity, Jennie was among many who donated food to the families of strikers from Monkton Coke Works.

When the National Union of Mineworkers eventually voted to end the strike, there was no settlement. Thatcher did not make one single concession. The year-long dispute had culminated in defeat for the strikers and heralded the collapse, not only of the UK's large-scale mining industry, but of many heavy industries across the country.

The subsequent disassembly of these sectors devastated many working-class towns, with the cities of northern England significantly impacted by Thatcher's deindustrialisation. Heavy industries had been the lifeblood of the communities that they sustained, with communities relying on this employment to support their families. The unions, despite their faults and limitations, had given the workers in these communities strength and solidarity, instilling a feeling of belonging in a shared working-class experience. The decline of coal mining, shipbuilding, heavy engineering, and manufacturing, notably in the North East, led to long-term, mass unemployment.

One of the casualties of the 1984 strikes was Fishburn Coke Works, in nearby Durham. It had produced high-grade sunbrite coke for the domestic market for decades but closed in 1984. The responsibility for fulfilling that quota fell to nearby Monkton Coke Works. Thus, production at the Hebburn plant continued to increase and switched from producing metallurgical coke predominantly for European customers, to domestic coke.

As the once-proud, formerly all-powerful coal industry approached its sad swansong, the incoming entropy of entrenched joblessness across the North East exacerbated a general feeling of rancour towards Thatcher. With the perspective of a broader lens in which to view this period, it is evident that the region's heavy industry decline had started before Thatcher came to power, due to poor governmental economic management and globalisation. However, Thatcherism certainly accelerated its decline, driving a deeper perception of a North-South class divide within the country, with a Conservative and rich South being anathema to a Labour-dominated North.

Another divisive programme that Thatcher introduced in the early part of the 1980s was her 'Right to Buy' scheme, which enabled tenants of council housing to buy the property they were living in for a large discount. Thatcher launched the housing sell-off because public housing clashed with her free-market principles and because her cash-strapped treasury needed the money from sales. She was keen to neuter the local council authorities, first, because she mostly considered them to be bastions of socialism, and second, so that her government could pocket the proceeds from the housing sell-off. The implementation of this programme was spearheaded by Thatcher's Secretary of State for the Environment, Michael Heseltine. With his mane of coiffed blonde hair and clipped upper-class accent, he made for a striking media performer. As head of the environment department, his position comprised several responsibilities, ranging from housing and planning to environmental protection. Despite being enthusiastically heralded by Heseltine as an opportunity for capital wealth to transfer from the state to the people, not everybody subscribed to the benefits of the 'Right to Buy'.

Although Jennie was a tenant who could now afford to buy her home, she did not want to, as she felt it was unfair. In her view, the council housing that had been built on Monkton Lane Estate was designed to help people on a low income. Council house ownership was paradoxical to the socialist Labour doctrine that she had held close throughout her life, and Heseltine's policy epitomised everything that she opposed. However, after her husband David's stroke there was a need for a downstairs shower room as he could not use the upstairs bath. Jennie organised with the council to undertake the necessary modifications, but shortly after the council's inspection it was declared too large for just two people. With the threat of being moved to a smaller premises, Jennie decided to purchase the home that she had lived in for over 25 years. She paid for the shower room modifications herself and now had peace of mind that they would live there for the remainder of their lives.

Jennie continued her upwards trajectory within the political world. Tyne and Wear was one of seven metropolitan county councils that covered the whole country. Jennie's efforts in representing her Tyneside town had been recognised by key decision makers across the county. By 1984 she had been elected as Vice-Chairman of Tyne and Wear County Council, with responsibility for a constituency that now included Tyneside,

Newcastle, Gateshead, and Sunderland. This appointment represented a huge leap in Jennie's political journey.

In a letter to the county council from the National Union of Mineworkers at the time there would no doubt have been a feeling of empathy of their plight from Jennie, and also some feeling of resonance towards her own neighbourhood's ordeal with Monkton Coke Works.

> The miners are desperate for their communities, and this desperation forces them to action. A society which seeks economic progress for material ends must not indifferently exact such human suffering on some for the sake of affluence of others. This government, whatever it says, seems to be determined to defeat the miners and thus treat workers as not part 'of us'. They also seem to be indifferent to poverty and powerlessness. Their financial measures consistently improve the lot of the already better off, while worsening that of the badly off.[81]

The following year, in March 1985, Jennie was voted in as the Chairman of Tyne and Wear County Council as the first ever female to be installed in the role.[82] It had been quite a journey for the 62-year-old, who was now a grandmother of eight. In the space of just over a decade, Jennie had gone from attending Wednesday evening meetings at the Women's Labour Party in Hebburn to now leading a predominantly male board that oversaw a collection of boroughs covering the whole county.

From her spacious office on the top floor of the Sandyford House headquarters, Jennie had panoramic views of the city and a secretary in the next room willing to help wherever possible. Her position would grant her access to dignitaries and directors, giving her on-the-job exposure to leaders from whom she could shape her own unique leadership style.[83]

Yet behind the honour of the role lay the same egalitarian Jennie. A journalist who wrote a feature on her described Jennie at the time:

> [As] a woman of character and fierce determination, couched in gentle grace, who asks for nothing more than people to accept her for what she is, and treat her accordingly. She is a woman of her word, and there are many people who could do worse than take a leaf out of her book.[84]

The same article reveals how Jennie was feeling about her newfound responsibilities.

> My job was to be a mother so I cooked and sewed and knitted and loved every minute of it and if we did without things then it didn't matter because we had a strong loving family and that's really all anyone needs. I'm just the same as I've always been. I'm not used to it at all [having a secretary]. I'm usually the one getting up and dashing round to get things for other people.

It's smashing being County Council Chairman, [...] but it's even better being a grandma. There's no denying that.[85]

— Jennie Shearan.

Jennie's role as chairman was multifaceted, with one key element being that of a figurehead for the county. When she met His Royal Highness, the Duke of Gloucester, during a two-day tour of Tyne and Wear, alongside her for the event was her daughter, Barbara, now 42 years old. During her year in office, Jennie was invited to the Queen's Garden Party, a summer event hosted at Buckingham Palace, and Barbara accompanied her then as well.

As Chairman of Tyne and Wear County Council, Jennie was allowed to choose a small gift to present to official visitors. She had a plate made with the coat of arms of the old Hebburn Urban District Council, a picture of Hebburn's Ellison Hall and a drawing of the famous Davy miner's safety lamp which was invented in her town. It was a memento that officials from as far as America to Zimbabwe received, with Jennie quoted as saying, 'I'm really proud to think a bit of Hebburn has gone all over the world'.[86]

Jennie tackled her engagements with enthusiasm and energy. She officiated the openings and launches of many new businesses across the county. At the openings of local train stations, she unveiled plaques with her name.[87] She was invited as guest of honour to important regional events such as the investiture of the Bishop of Durham. When the results, participant times and photos were published for the Great North Run, one of the biggest running events in the world, it was Jennie who wrote the Preface to the book that was subsequently circulated. Jennie was called on to officially open everything from the region's Folkmoot Music Festival, celebrating folk musicians and dancers from around the world, to the opening of a new section of the Bowes Railway Heritage Museum.

As the first woman to become Tyne and Wear's chairman, Jennie felt she could bring a new dimension to the position, helping and involving women where possible. Only a few weeks into her chairmanship, Jennie organised a luncheon for the female staff at the headquarters and for the women councillors. No one had ever thought of this before, and it was a resounding success.

> I have different ideas to the men, just little things and slightly different approaches. It's really only doing the job with a woman's point of view.
>
> I'm thoroughly enjoying myself. It's hard work but you get the chance to meet the top people and the ordinary rank and file and you can't do better than that, because people generally are the most important thing. It's really a joy and doesn't seem like work at all.[88]
>
> — Jennie Shearan.

Her husband David was very proud of everything that she had achieved and fully supportive throughout her time as chairman. However, he was not physically well enough to join Jennie on the many dinner events that she had to attend. In his place, Jennie would often be accompanied by Father Dolan, a Catholic priest in Hebburn who had officiated the marriages of some of her children and christenings of some of her grandchildren, and with whom she had developed a close friendship. The role was also teaching her about the full extent of governing bodies that had a say in every element of the management of the county. Meeting minutes from the time reveal Jennie discussing local issues with the heads of the Economic Development Committee, the Waste Disposal Committee, the Arts Council, and the Policy and Resources Committee. From the Planning Committee to the Public Transport Committee, Jennie was dealing with issues ranging from land reclamation to approvals of grants, handling large budgets, and shaping the future of the county.

Unfortunately for county councils across the country, during the Thatcher years, local government in England was markedly restructured. By 1986, Jennie was to be the last ever Chairman of Tyne and Wear County Council, when Thatcher abolished these administrations and transferred their powers to individual cities or unelected bodies attached to central government departments. The official aim of these reforms was to improve the efficiency of public service provision, but there was also an underlying political element. Thatcher was seeking to diminish the political influence of the metropolitan county councils. Greater London's county council had become a thorn in Thatcher's side, with a series of policies that countered her vision for the country being mooted. On top of that, the major conurbations of the North were Labour strongholds where councillors frequently blocked her application of government policy.[89] Just like the trade unions, the county councils were threats that had to be eliminated.

The reforms spelt the end of Jennie's official involvement with the council. She had learned so much from her time there, and now it was time to put it all into practice for the issue that was closest to her heart. The matter of Monkton Coke Works was going to continue to deteriorate the quality of life of her neighbours for as long as the plant was profitable for its owners. Jennie had learned from her time in politics that the ability of the council to drive an improvement in this matter was limited. Yes, Jennie had secured rent rebates for the residents living in the immediate shadow of the plant, but nothing had changed with regard to the ongoing pollution that the site created.

> I managed to get a rate reduction for people on the housing estate so that they can clean the muck out of their houses, but you cannot clean the muck off your chest. There are playing fields less than one hundred yards from the works, but the children cannot use them because they get so filthy. Hebburn used to pride itself on being the first town in the country to be a one hundred per cent Smokeless Zone, yet Monkton is allowed to get away with this.[90]

— Jennie Shearan

Across May and June of 1986 there were particularly bad incidents of pollution in Monkton Lane Estate, with homes and cars coated in a yellow slime, and 80 mile an hour winds scattering coal dust across the neighbourhood. Jennie was now an established figure in the local press, and expressed her disdain in extensive interviews on the matter. She was not alone in her fury. Councillor McTo-minay described the plant as a 'scourge', adding, 'We know what has happened to trees and flowers up there. If it can kill them off, what's it doing to people?'[91] Councillor Bamford echoed the community's anger by suggesting that, 'We should be looking at the possibility that Monkton Coke Works is putting so much pollution into the area and affecting the people to such an extent that it shouldn't be there'.[92] While the councillors' words were supportive, no tangible improvements were made.

It had become clear to Jennie that if the neighbourhood was to have any possibility of getting the plant cleaned up, the residents themselves needed direct and collective action. The issue would have to be addressed directly with National Smokeless Fuels and its parent company British Coal, recently renamed from the National Coal Board. The issue also would have to be shared with the government's Inspectorate of Pollution. To approach the problem in a structured manner, Jennie recognised that she was going to have to get herself organised and systemically empower the people of Monkton Lane Estate.

Her realisation proved timely, because in August 1986 National Smokeless Fuels put forward an outline planning application for the construction of a power generation station at Monkton Coke Works.[93] It seemed as though Goliath was continuing his march forward, with David helpless to intervene. The proposal received by the Town Development Sub-Committee of South Tyneside summarised that excess coke oven gas could be utilised to generate electricity, rather than simply being flared away.[94]

British Coal's vision was that the energy produced could satisfy the site's own electricity requirements, and the forecasted thirty per cent surplus could be exported to the North Eastern Electricity Board. As part of their idea, they would need to build a gas turbine and a chimney to satisfactorily disperse the elements of combustion. The proposed chimney would be 55 metres in height, as tall as the Leaning Tower of Pisa, which was the upper limit of what the Inspectorate of Pollution allowed for any chimney at the time.[95]

The firm's plan was an economically savvy way to convert the otherwise wasted gas into valuable electricity for their own means, and monetise the remainder. National Smokeless Fuels were careful to outline the steps that they were taking to ensure the process would be closely controlled, and they even argued that the benefit of the initiative would be that the infamous hellish flaring would be reduced. Their application also notified the council that the European Commission had provisionally granted funding for the project, amounting to one third of the total expenditure, and that a decision had to be made quickly in order to meet the grant's deadline.

Tellingly, the proposal failed to properly address the plant's poor track record of controlling pollution and the impact that this had had on the surrounding residential areas. It also failed to acknowledge that any community trust around compliance to the requirements of the Inspectorate of Pollution had eroded after years of horrendous pollution from the plant.

Due to the height of the proposed 55 metre chimney, National Smokeless Fuels were obliged by the Town and Country Planning Act to advertise the application to the local press. When the announcement appeared in the *Gazette* in February 1987, South Tyneside Council's planning department distributed consultation letters to over 200 residents in the surrounding areas. The notification sparked immediate and vehement opposition.

This was Jennie's praxeological watershed. She knew that this was the time to incorporate everything she had learned, and galvanise her community. A game of chess had been set in motion, and she was determined that the residents would no longer be mere pawns. Jennie decided to form a committee of her own, dedicated exclusively to the matter that affected so many people in the area. The express purpose of the task force would be to harness the community's aversion to further development at the plant and raise awareness about the level of pollution that the site was creating.

Throughout her time on the council, Jennie would often say, 'You can move mountains with people power'. She had been politically engaged her whole life, and knew that when the will of the people manifested itself into one cohesive entity, the ability to fight an issue increased exponentially. The lone voices of the residents of Monkton Lane Estate would be marginalised no longer. As nobody was prioritising their interests, they were going to have to take matters into their own hands.

In the same month that the *Gazette* printed the press notice, Jennie, now aged 65, started bringing together the people that were closest to her, and who best understood the daily realities of the issue. This team of neighbours were living and breathing the situation every day and could represent the whole area as one voice. The residents of Monkton Lane Estate were all too familiar with the very real proposition of being denied fair and meaningful participation in the decision-making that would shape their lives. This was their time to take responsibility for cleaning up the plant on their doorstep.

Jennie would act as chairman and her eldest daughter Barbara would be vice-chairman. Barbara still lived in Hebburn and had grown up witnessing the daily effects of Monkton Coke Works. She knew that what they were about to embark on would be time-consuming and stressful, but there was no doubt in her mind that she would stand shoulder to shoulder with her mother on this matter. Marian Carey was a neighbour and good friend, and she joined too. Jenny Johnson, whose husband Bill had long since retired from his mobile business supplying groceries to the estate, also committed herself to the cause. Other long-time residents, Maudy Cork and Jenny Lowry, also joined, and they would act as treasurer and secretary respectively. These women were mothers, and some were already grandmothers. They were all modest homemakers who had lived most of their life on Monkton Lane Estate. They were thoughtful and articulate, and they were ready to link together and fight for their community.

The nucleus had been formed. Other members, including the core group's husbands, would follow along the way, and also help strategise on how to clean up their estate. They had a lot going against them in their battle to take on as immense an institution as British Coal. They were not scholars. They had no access to finances or even office equipment. They were middle-aged Geordie housewives trying to confront a male-dominated industry that was managed hundreds of miles from their home. They were going to need to be irrepressible and innovative if they even had a chance of being heard.

CHAPTER FIVE

HEAD TO HEAD

'Welcome to Jenny Shearan country.'

In the following month after National Smokeless Fuel's press release was published, a public meeting for the local neighbourhood was called, so that the Town Development Sub-Committee of South Tyneside could garner an understanding of the residents' concerns around the proposed application. This would be the first time the populace could meet the key decision makers, and attendance was extremely high, with over 250 locals present.

Within the packed hall at Luke's Lane Junior School, where the meeting was hosted, was Jennie's newly formed committee. For the first time, the people of Hebburn were coming face to face with their Local Planning Authority, the manager of Monkton Coke Works, and representatives from National Smokeless Fuels. It was finally time to get their voices heard by the men who were making the decisions.

Jennie's committee came prepared. They had researched what negative effects the installation of the proposed generator would have on the community and each had shared what they learned in written letters of objection to the council. This was the pre-internet era, with a writing pad in place of email, and the local library in place of Google. Following extensive investigation, the committee had located a team of experts in Norway who had studied the impact of such a generator on the community. Their data pointed to an increase in pollution, notably in the form of acid rain.

Adamant that no development plans should be discussed until the existing pollution issues were investigated, the residents left National Smokeless Fuels in no doubt about their feelings. It was a fiercely debated meeting, lasting over two hours. Several times during the assembly, the councillor who chaired the meeting had to call for order. The residents bombarded the representatives with a barrage of complaints about unacceptable filth coming from the plant. They listed dirt, noise, sleepless nights, odours, traffic hazards, and increased emissions as the main reasons why the size of the plant should reduce, not expand.

A local farmer who owned land around Monkton Coke Works described the damage to his crops. One resident recounted all the previous broken promises about keeping pollution under control. Jennie asked why permission had been granted to double the number of coke ovens from 33 to 66 and called for a survey to be done of pollution levels from the plant. A mother explained how her daughters' toys were now discoloured by the smoke from the coke works. Another mother told how she left the back door open while she was baking pies and returned to see them ruined after a fine line of dust had settled over them.[96]

Beyond the anecdotes of the constant battle to keep abreast of the filth, multiple residents insisted that National Smokeless Fuels should spend their money on filters for the existing system. In support of the residents' concerns, one councillor said that British Coal was a 'law unto themselves. I do not trust them. If this plan gets the go-head the trouble and dust will be threefold'.[97]

Within the overwhelming expressions of discontent, there was a sign of future community tension to come. Jennie was quoted as saying, 'We should stand up as one and say no to this plan', but one worker at the plant exclaimed to the residents, 'You supported us during the Miners' Strike, now you are asking for people to close the plant'. The residents were not opposed to employment for the people of their town. They understood the economic vulnerability of the area. They were simply exasperated by the damage that the plant was causing to their community.

In March 1987, after digesting the information that the residents had shared with the Town Development Sub-Committee, the council refused British Coal's application for planning permission for the power generation station, with a unanimous vote.

This was a pivotal milestone for Jennie. It was a reminder that when a group of people unite, they can be heard. Over a timespan of decades, the country's nationalised Coal Board had added to their Hebburn plant at will, with a total of 66 coke ovens now operating around the clock, all year long. Now, the environmental discrimination that the residents had been made to tolerate for so long had at last been recognised and an extension to the plant had been vetoed. It felt like the beginning of a sea change for the residents of Monkton Lane Estate.

Jennie saw this as a key opportunity for her committee to mobilise the whole community into forcing the works to clean up its act. Her close-knit assemblage needed a simple and memorable title, to build recognition and deepen participation. They were officially named the Hebburn Residents' Action Group. With Jennie's political experience and far-reaching connections, she was the natural leader of the group, but this effort was nothing without the people alongside her. This was not Jennie's cause. This was not even the cause of the other members. It was the cause for the whole community.

Just as the Women's Labour Party had set aside Wednesday evenings as the recurring time for their members to meet all those years ago, Jennie hosted the Hebburn Residents' Action Group each week at her home. Once the tea had been poured, milk and sugar added, and everyone had chosen a biscuit from the tin, the embattled residents began formulating a plan.

It was during these early meetings that the group crystallised their mission to improve the living conditions and the health of the neighbourhood. With their goal defined, they acknowledged that if this was to be achieved, they would have to devise a comprehensive programme to pressure the local council and government watchdogs to stymie any future expansion plans and clean up the site. Generating mass awareness of the increasingly polluting plant was key.

Jennie understood the need to rouse the residents of south Hebburn and get their written input on this issue. Soon after the heated public meeting, she collected a petition, bearing signatures throughout Monkton Lane and Luke's Lane Estates that opposed the development. In the space of 36 hours, the

Hebburn Residents' Action Group collected 1,000 signatures. While visiting each house door-to-door, the horrifying gravity and urgency of the problem were revealed. Jennie saw countless residents suffering from severe breathing disorders, particularly on Hexham Avenue. With many neighbours needing oxygen support in their homes, frequently using inhalers, or suffering from sore throats, she saw a community condemned to a life of almost constant illness by breathing in coal dust and toxic emissions. The petition's message to British Coal was simple: clean up Monkton Coke Works before any extension to the plant is even considered.

With Jennie's close connections to the local council and local press, she organised a journalist to document her presentation of the petition to Councillor Gerry Graham. The photo in the next day's newspaper, with Gerry and Jennie jointly holding the petition, and her daughter Barbara by Gerry's side, is the earliest recorded reference of the Hebburn Residents' Action Group. It was to be the beginning of a blusterous relationship with both the press and the local politicians on this matter.

> We are strongly opposed to any further develop-
> ments of Monkton Coke Works until the pollution
> is sorted out.[98]
>
> — Jennie Shearan.

Having witnessed the harrowing effects of the plant on so many neighbours, Jennie knew that a detailed investigation was going to be needed. She resolved to find a way to establish a correlation between the plant's pollution and the bronchial complaints around the community. Jennie started to push her local council contacts to commission a health survey on the residents surrounding Monkton Coke Works.

Jennie also began gathering evidence of the pollution on video and, with the help of a local film editor, a short film was created with the footage that she had accumulated. Jennie premiered her documentary at a special exhibition, dubbing it *Certificate 18 for Anyone Interested in Fresh Air and a Clean Environment*. Naturally, the local press were invited to the low-key première.

> Everything in the film was shot from my front door. We decided to get a record of what it is really like when the plant is working at full blast, throwing out all the filth and fumes we have to breathe. This is what the residents here have to put up with and I think anyone who sees the film will understand why we are so angry and unhappy.[99]

— Jennie Shearan.

While pleased with the local press coverage, the action group needed to get publicity on as broad a level as possible about the issue that had plagued them for decades. They were desperate to halt the growth of a plant that was continuously threatening to expand. The coke works that overshadowed their homes had subjected their community to horrendous environmental degradation, and was on the precipice of worsening. They set about writing frequent letters of protest about the pollution to local and European Members of Parliament. They even wrote a letter to the Queen, with a request for support. Buckingham Palace's reply helpfully pointed them in the right direction of the ultimate decision maker.

> I am commanded by The Queen to acknowledge your recent letter with which you forwarded a petition from the Residents' Action Group signed by people living in Hebburn and the surrounding districts. In accordance with constitutional practice and at Her Majesty's direction, this petition has been forwarded to the Secretary of State for the Environment.[100]

From this correspondence, the action group now had a clear understanding as to who they ultimately would need to escalate the issue. They now needed to secure significant press coverage of the problem, and an event to which Jennie could invite her local media contacts.

The action group knew that this issue derived its power from the affected communities uniting in their cause. As such, they sought to engage all the people who had been affected. With community sentiment against Monkton Coke Works at a high, the action group called for a peaceful protest outside the plant in May 1987. Lending from the optics of the recent Miners' Strike pickets, while being careful to not summon the Miners' Strike's climate of insurgency, the women created a multitude of placards and posters for the demonstration. There was mass participation for the event, for which capitalised slogans were produced and carried, as 200 residents marched around the plant. 'Acid rain.' 'No more pollution.' 'Enough is enough.' 'We want clean air.' 'Clean it up.' 'Pollution and smell.' While they walked through the estate towards the plant, they chanted, 'What do we want? CLEAN AIR! When do we want it? NOW! What have we got? MUCK! What do we want? CLEAN AIR!'[101]

The local press were there to greet the demonstrators once they arrived outside Monkton Coke Works. Barbara was pictured in the local press the next day, among a throng of fellow protesters, holding the plea, 'LET US BREATHE', with Jenny Johnson not far behind her, showing the satirical statement, 'THIS IS 100% SMOKELESS ZONE'. At the front of the line of protesters was Jennie, holding the placard that the group had made for her: 'WELCOME TO JENNY SHEARAN COUNTRY'.[102]

The slogan was a gentle play on the tagline 'Welcome to Catherine Cookson country', which had been propagated by Tyneside's tourism board. Born and raised in east Jarrow, Cookson had become an established author, with global sales of her novels topping 100 million. Her books were inspired by her deprived youth, and almost exclusively focused on South Tyneside. Such was her popularity throughout the country that local tourist officials decided to rename the borough, 'Catherine Cookson Country'.

In the eyes of her fellow activists, Jennie was to Hebburn what Catherine Cookson was to Tyneside. Regardless of how Jennie's name was spelt on the placard she held, the message was clear. The protesters wanted clean air. They were fighting for equal access to the decision-making process about the future of the plant, and the future of their community, so that they could live in a healthy environment.

From this moment onwards, the issue was no longer to be discussed solely within the pages of the *Gazette*. Coverage of the protest now reached the newspaper that covered the whole of the North East of England, and Jennie was able to voice the neighbourhood's growing resentment about the conditions in which they lived to a much broader audience beyond South Tyneside.

> We've had this for 30 years, but nobody will do anything about it. Something drastic needs to be done because the problem is getting worse. The council has six schools here which are the future of the area. If you got some peace from the noise, smell, and dust it would not be so bad, but it is here all the time. We are fighting for our environment and will not give in. They need to clean up their act or close it down. No one will take any notice of us, and this is why we are determined to clean up our environment. Come hell or high water, it will be done. We have breathed in enough fumes, acid rain, and pollution. You name it, we've got it, and to me that's not on.[103]
>
> — Jennie Shearan.

Meanwhile, National Smokeless Fuels refused to accept the decision made by South Tyneside's Town Development Sub-Committee. They were resolute in setting up a turbine to convert gas into electricity, as a means to create incremental income for the plant.

In June 1987, the firm put forward a second application to the Local Planning Authority, again advertising it in the *Gazette*. This time, they supported their proposal with the announcement of a clean-up scheme costing over £400,000. Landscaping and tree planting were introduced.[104] Efforts to reduce problems caused by dust were set in motion, including concreting over more areas, erecting windbreak fencing, providing a permanent water sprayer, building new access roads and a land bridge, and sealing coal stockpiles with latex spraying.

The community immediately saw the gesture for what it was; tokenism.[105] Representing the community's scepticism, Jennie dismissed these proposals as merely cosmetic which would not tackle the main problem of reducing the pollution from the ovens.[106]

> We accept the company has to run its business in the best way it can and we know the real blame lies with the people who allowed the plant to grow so big with two housing estates on its doorstep. But we want to be able to breathe fresh air and nothing that has been done will make that possible. We want the coke ovens cleaned up and some effective action taken to cut down the amount of pollution coming from the plant.[107]

— Jennie Shearan.

National Smokeless Fuels was at loggerheads with the Hebburn Residents' Action Group. Roy Howson, the Managing Director of National Smokeless Fuels, summed up the mutually conflicting dilemma in which both parties found themselves.

Every coking plant is a polluter. It's the nature of the process. My job as manager of the plant is to reduce that to a minimum. New technology enables us to do that. We have embarked on many new things over the years. A new by-product plant, better loading, better streaming. Everything's aimed at reducing pollution. It would be impossible to eliminate it completely.

The money has to be spent to enable it to operate profitably in the future. We've got to have a plant, which means we've got to live with the local people. If we don't have a plant, we will fail.[108]

— Roy Howson.

While Howson sympathised with the residents, in the battle between healthy people and healthy profits, it was clear that profits were paramount.

National Smokeless Fuels launched an official appeal to the Department of the Environment against the council's refusal to grant planning permission. To find a solution to the matter, a public inquiry was scheduled to take place at South Shields Town Hall in December 1987.

In the UK, when there are contentious issues around land use developments that need to be resolved, the government's Planning Inspectorate will hold an official public review of evidence from both sides. The planning inspector is the judge at a public inquiry, and acts as the interface between the public and the land use planning system. The inspector's role is to impartially hear all the evidence, which can come in the form of written representations or hearings. Over the course of several days, the inspector listens to perspectives from all sides of the argument. The following weeks are spent digesting the information and distilling the salient points into a report that comes with a recommendation, once the inspector has made up their mind on the outcome of the dispute. The inspector then takes the verdict to the Secretary of State for a final decision, who rarely disagrees with the inspector's recommendation.

The government typically only allows a small portion of requested public inquiries to take place, due to the expense involved with running them. The cases that are chosen are generally based on how high profile they are and the extent of the media coverage on the topic. Thus, the years of Jennie getting the matter discussed in local newspapers had helped to get this issue addressed at the highest level in the country.

After so much effort, the action group had won the right to be heard at a national level. While they had a strong case and plenty of materials to support their position, it was critical that they secure professional representation. Since the action group's conception, they had exercised financial prudence, but the fees required exceeded the little money they had available. The overall objective of a public inquiry is to win your case by proving that your position, unlike that of the opposition, is supported by government policy and the broader objectives of society. Jennie had the evidence but she did not have legal qualifications. Thanks to the well-publicised history of the issue, local solicitor Ian Bynoe was aware of the case. He stepped in and generously offered to waive his fee for his services. For the campaigners, this level of pro bono legal support was much needed. The action group's well-organised filing system for the material that they had gathered helped Ian prepare a well-focused case in the months leading up to the inquiry.

For one of the sessions in the inquiry, the inspector had agreed to listen directly to the residents' concerns. A special two-hour meeting was scheduled, to be hosted at South Tyneside College in Mill Lane, with Monkton Coke Works in the background. The Hebburn Residents' Action Group knew that strong attendance would not only enable the community to voice their concerns but also demonstrate to the decision makers the depth of public concern.

The action group set about maximising participation and developed a tri-fold leaflet that was delivered by hand to thousands of households in the area, urging them to both show up to the special residents' meeting and be present throughout the public inquiry. In the mid-1980s, the singer-songwriter Sting was at the height of his popularity and his signature song at the time was *Every Breath You Take*. The musician was famous worldwide and was especially admired in his hometown of Wallsend, which is just on the opposite side of the River Tyne to Hebburn. The lead headlines across the action group's leaflet were an acerbic twist on his most played track, with the leaflet leading with 'Every breath we take ...' and continuing onto the back page with '... Every move we've made'. Jenny Johnson was in the front photo of the leaflet among fellow protesters, as they marched towards Monkton Coke Works. It was a well-thought out and well-executed piece of publicity. Among the details of their campaigning so far, and the history of the plant, was Jennie's name and address and telephone number, prominently placed as the sole point of contact for the action group. Incorporating professionally shot images from their earlier protest, and photos of the polluting facility in full flow, their call to action worked. Resident attendance was high throughout the public inquiry.

Expert witnesses from every field associated with the proposal were brought in to offer their perspectives at South Shields Town Hall. A meteorologist disclosed his opinions on the impact of a power generator on pollution levels. An acoustic engineer was consulted on the potential noise level increases from the gas turbine. Doctors were invited to share case studies of bronchial asthma caused by sulphur dioxide.

The argument that the appellant, National Smokeless Fuels, led with focused on a list of benefits that the proposed power generator would bring to the community. An important thrust of their reasoning was that the financial profits generated from expenditure savings and export earnings would help secure continued employment for the present workforce. A key undercurrent that ran throughout their case was that the long-term viability of the plant would be put in doubt if the power station project was blocked. They positioned themselves as a valuable contributor to the local economy, employing over 200 people with a salary base of £3 million, and paying South Tyneside Council just under £300,000 in rates.

National Smokeless Fuels also explained that the £700,000 in annual electricity savings would help fund continuing efforts to reduce pollution from the plant. The firm outlined a commitment to capital expenditure towards ongoing maintenance of the site. They cited that across the 1970s they had invested in the sequential charging of all ovens in a bid to reduce emissions during charging, at a high cost to the firm. They also explained that they had introduced automated oven frame cleaning equipment to reduce smoke leakage. However, within this list of equipment, the elusive, and much-needed, grit arrestor panels were conspicuous in their absence.

In addition, the firm argued that the waste gas that would be used to produce the electricity would save the equivalent of 10,000 tonnes of coal being burned elsewhere each year. Thus, by deploying the gas at the plant, instead of wasting it, there would be a net reduction in the amount of sulphur dioxide and other gases emitted into the atmosphere.[109]

Finally, they acknowledged the pollution that the plant had historically caused but pointed out that the Smokeless Zone initiative that the government had introduced in 1956 related only to domestic properties. This, perhaps unintentionally, articulated the crux of the issue that both the firm and the residents were enduring. A coking plant and a housing estate created an incompatible dialectic, no matter how expertly National Smokeless Fuels portrayed themselves, or what efforts they made to curtail the pollution that besieged the community.

It was now time for the Local Planning Authority to share its perspective. Due to the onslaught of residents' complaints, the council recognised that their previous laissez-faire approach to planning decisions in the area had to change.[110]

Their newfound more discerning perspective had also been triggered by a recent directive from the Department of the Environment. Within the memo, the department had provided advice to all planning authorities dealing with applications involving hazardous industry.[111] As such, the council viewed National Smokeless Fuels' planning application from several angles.

Their first issue was that there was still no grit arrestor panel in the proposal, even though this had been a commitment for the newest batch of oven batteries that had been built in 1980.

Delving into what implications a power generator may have on the local community, the council also expressed concerns about fire explosions in the event of a gas leak and noise breakouts from the increased activity.

Another major issue was National Smokeless Fuel's failure to propose the installation of flue gas desulphurisation apparatus to counter the pollution that the power generator would create. Flue gas is the combustion gas exiting into the atmosphere via a flue, which is a pipe used for conveying exhaust gases from an oven. Armed with the action group's research findings, the council stated that the lack of provision for the removal of sulphur dioxide and other harmful emissions from the proposed chimney would lead to an increased possibility of acid rain. Desulphurisation equipment was thus deemed necessary to treat the gases at the plant and eliminate pollution.

The council also analysed recent efforts taken by National Smokeless Fuels to clean up the plant. They determined that the firm's remedial attempts to counter the pollution had only been undertaken after a catalogue of protestations from the residents. This indicated to the council that the firm had not done all that would have been expected to contain the outbreaks of heavy pollution.

What sealed the fate of the council's position on this matter though, was the number of grievances that they had received about the plant. In the previous year, their public health department had received 51 complaints relating to dust, odour, smoke, noise, polluted soil, and flooding to the highway. The notorious reputation of the plant had been palpably exposed by the residents' complaints, casting doubt on the ability of National Smokeless Fuels to effectively construct and operate their proposed power generation station without causing further distress and disruption. The extent of the resistance had helped the council see through the firm's rhetoric around economic benefits and insinuations around possible job loss.

Reference to the clause 'best practicable means' was introduced at this stage in the case. This clause had been at the cornerstone of the UK's system of regulation of industrial air pollution for over a century. National Smokeless Fuels were required to follow best practicable means to control pollution from the plant. It was a strict set of criteria on which performance and decision-making were judged. It comprised the firm's technical knowledge, the financial implications of new measures, and the overall circumstances under which they were operating. The council argued that the plant had not followed best practicable means, and the history of management of the plant was so unsatisfactory that it did not bode well for any further developments. As the firm had experienced difficulties operating a coke works which employed long-established technologies, it was asserted that they could not be expected to have the level of expertise deemed necessary for the safe and satisfactory operation of the proposed gas turbines.

The council concluded that while there may have been benefits in the proposal relating to the utilisation of a gas resource that would have otherwise been wasted, they considered that this was outweighed by the likely effect of the proposal in terms of pollution and noise. They deemed the operation of the existing plant as relevant in that it demonstrated the inability of National Smokeless Fuels to operate the plant to the required amenity standards. In the eyes of the council, this inability could easily extend to the operation of the proposed power generation station. They voted to reject the application based on the premise that incorporating a gas turbine that discharged into a 55 metre chimney would only lead to further damage to the environment. By citing National Smokeless Fuel's poor record of controlling pollution in the past, and their fears of future pollution from the works, the council publicly accepted what the Hebburn Residents' Action Group had been saying all along: that the coke works needed to be cleaned up.[112]

It was finally the turn of the residents to address their concerns.

In the lead up to the public inquiry, the letters that they had sent to the inspector had been photocopied and incorporated into the dossiers of the key decision makers at the inquiry. The collection unequivocally captured the level of opposition to the proposed new development and gave the investigators a robust understanding of the residents' sentiments towards the Byzantine plant. The most frequently cited complaint was about the pollution caused by the existing operations which would be exacerbated by further development on the site. The residents' accounts painted a miserable picture of life next to a coking plant which constantly compromised living standards and made it 'hell on Earth'[113] for the people living there. They detailed the 'dreadful dust'[114] that pervaded even the food cupboards. They described the polluted water that had seeped through to the surrounding ground and killed plants. They called for an investigation into the damage done to the health of nearby residents before any expansion scheme could be considered. The missives also expressed how incongruous it was that Hebburn had been designated a Smokeless Zone.

> The site is already extensive and somehow has been allowed to creep steadily over the years. I have lived in Jarrow for nearly fourteen years and Monkton has never been as large as present. The only time life was clean was during the year-long Miners' Strike when the plant was idle. I was under the impression that we lived in a Smokeless Zone but having witnessed clouds of grey and black smoke pouring from this site, and more noticeably under cover of darkness, I do not think this must be the case. We have explained many times and have been told by people not living underneath all this dirt that they can't see anything wrong. They should come and see the black smoke that is pouring out of the coke ovens when the ovens are being pushed.[115]

The residents now had a chance to voice their concerns directly to the inspector. Those in attendance from Monkton Lane Estate may have been slightly daunted by the oppressive, official atmosphere of a public inquiry. There may have been a feeling of trepidation when facing the intimidating lawyers representing National Smokeless Fuels and the council bureaucrats that flanked them. Nevertheless, for the residents, this matter was not about a profit-driving exercise; it was about their lives. Barbara, speaking on behalf of her community, stood up and gave her testimony to the inspector.

> Thank you for giving me this opportunity to voice my objections to any further developments at Monkton Coke Works. Over the years this plant has been allowed to grow and grow out of all proportion to the detriment of the residents living close by. Our lives and environment have become intolerable because of the effect this plant has on our everyday lives. It is in operation 24 hours a day, seven days a week, 52 weeks of the year.
>
> The effect of this is filth, muck, pollution, and noise. You have to live with the conditions to know how it can get you down. You can't put the washing out because it gets covered in grit. You can't sit in the garden because when the ovens are opened you are hit with spots of moisture which sometimes sting your skin. The smell which comes from the plant seeps everywhere.
>
> My parents have lived directly facing the works for over 30 years. When we first moved here it was nowhere near as large. Now it has grown three times in size and the ovens have doubled from 33 to 66. This expansion has made life unbearable for the residents living nearby.
>
> The National Coal Board won't spend money on an extractor machine which would have helped the

amount of filth and pollution the residents have had to put up with over the years.

We appreciate that the site manager has a job to do. The way I see it, his only priority is meeting the target that the National Coal Board has set him, even if it means workmen pushing the ovens when the coke is not cooked and all that filth is let out into the atmosphere.

I am appalled at the way the works have been allowed to grow and I think the time has come for the residents' complaints to be heard.

We demand a cleaner environment in which to live. I am hoping this time our views will be heard.[116]

— Barbara Burns.

Jennie was next and shared similar opinions to that of her daughter. While both Jennie and Barbara were able to persuasively elucidate their feelings to the inspector, they also wanted to give the attendees at the inquiry a sense of what life was like on the estate, beyond the use of words. Using a cassette player that she had brought with her to the venue, Jennie played tape recordings of the disruptive clanking from the plant at night. This was a potent evocation of what night times were like for the nearby residents, but Jennie had something even more compelling up her sleeve. She wanted the inspector to see and smell the neighbourhood. Running contrary to the staid conventions of the inquiry, she brought along blackened linen that had been soiled by the plant's sulphurous fallout as it hung out to dry in her garden. Unfurling the dirty dust-covered fabric from her bag, the overpoweringly malodorous discoloured cloth attacked the senses of the panel. The public inquiry may have been held several miles from Monkton Lane Estate, but Jennie had brought Monkton Lane Estate to them.

Over the course of five days, several million words of evidence were heard from National Smokeless Fuels, South Tyneside Council, and the Hebburn Residents' Action Group. With the abundance of evidence that had to be analysed and debated by highly paid experts, the cost of the inquiry had totalled over £50,000, the equivalent of £150,000 in today's money.

It was time for the inspector to give his final words, before adjourning the inquiry. Reiterating the council's attestation that a high standard of housekeeping had not been maintained at the plant, and was unlikely to be in the future, the inspector described the plant as a pollution blackspot. He came to the irrefutable conclusion that planning permission should be refused.

Following the protocol of public inquiries, the inspector wrote up a summary of his findings in a report. It was now left to Nicholas Ridley, the government's incumbent Secretary of State for the Environment, to consider the inquiry report.[117] The litany of complaints from the residents and the lack of support for the plan from the town hall chiefs had made the inspector's conclusion clear cut. The secretary agreed with the inspector. Ridley chose to prioritise the protection of human health and general well-being, the amenities of surrounding local residents, and the environment. The firm's request for a permit to construct a power generation station, gas turbine, and chimney was refused.

National Smokeless Fuels had completely underestimated the Hebburn Residents' Action Group. The activists had put forward a powerful case that they were victims of environmental injustice. Their vivid accounts of how the reduced air quality and the unabating noise had affected their lives played a key role in Ridley's decision to block the expansion of the coke works. It was a major victory for the overjoyed residents, and they duly organised a street party to celebrate. They had established themselves as environmental activists, and their efforts were gaining real momentum.

However, for Jennie, the fight was a long way from over. The plant remained in place and was as dirty and destructive as ever. The day after Ridley ruled in favour of rejecting the planning application, an interview with Jennie summed up the action group's pledge to continue the fight.

> We took on the big man and won, after we were told that we couldn't beat the big industries. This is a victory for all the people in the Hebburn Residents' Action Group and the people in the area. It's absolutely great and we'll be having a big celebration here. We won't stop until the air around here is clean. We're fighting for the health and the environment of the future generations of this town. We want to get rid of the smog and muck and dirt which is always around here.[118]
>
> — Jennie Shearan.

It was time to take this grassroots movement to a national level.

CHAPTER SIX

ELVIS HAS LEFT THE BUILDING

'So, where would you expect to put such a belching monster? Surely you would have been mad to build a couple of housing estates around it?'

In the 1980s, before the internet became the globally embraced communication tool that it is today, there existed several powerful media outlets in the UK that disseminated messages of all kinds to a diverse number of people. Print media in the country reached its highest ever newspaper circulation levels between 1986 and 1987.[119] Commensurately, the top three most-watched television broadcasts in the country's history took place in this two-year period. Newspapers and television were at their peak of consumption and were firmly ensconced in the daily lives of the country's populace. These media outlets fulfilled many purposes: they educated and provided information; they were a form of escapism; they served as public forums for the discussion of important issues; and they also acted as a watchdog for government, business, and other institutions.

The media theorist Marshall McLuhan famously coined the phrase, 'the medium is the message'.[120] By this, he meant that every medium delivers information and ideas in a different way, and the nature of the content is shaped and emphasised by that specific medium of transmission. Therefore, whilst both newspapers and television are suitable for the mass distribution of information, they are dynamic in different ways. For example, the 'Letters to the Editor' section in a newspaper allows readers to respond in depth to journalists, or voice their opinions on issues of the day, facilitating an exchange of opinions that television does not so easily allow. Conversely, television has the advantage of providing moving images, making a story come alive more vividly with a strong audio-visual component. Thus, the form of a message, whether it be through printed words or video footage, determines the way in which that narrative is perceived, and can substantially alter the way in which the audience understands the information being conveyed.

The television media landscape in 1980s Britain was dominated by two channels. The first was BBC One, part of a public broadcasting company, and the other was ITV, a commercial television station. This was a decade when over 30 million viewers, more than half the country's population at the time, tuned in to watch Den serve Angie divorce papers in the BBC soap opera, *EastEnders*.[121] For context, today, in the internet-dominated 2020s, the best-loved television programmes only generate half that amount of viewership. The most popular shows in the 1980s were soap operas such as *EastEnders* and *Coronation Street*, and sitcoms like *Only Fools and Horses* and *Auf Wiedersehen Pet*. Beyond the evening news broadcasts, topical matters were discussed in a variety of ways, ranging from the satirical puppet show *Spitting Image*, with its grotesque caricatures of public figures, to *Watchdog*, which aimed to investigate complaints made by viewers.

Since its inception, the activism that the Hebburn Residents' Action Group had been engaged in had been steadily picked up throughout the North East's media outlets. Having started as an issue that would be featured sporadically on the inside pages of the *Gazette*, with a daily circulation of just over 15,000, the matter had now become front-page news for the local newspaper. Moreover, the bigger publications in the region, such as the *Evening Chronicle*, and its morning counterpart *The Journal*, focusing on news north of the River Tyne and with circulations of more than double that of the *Gazette*, were regularly updating their readership on developments. South of the River Tyne, the *Sunderland Echo* had become equally prolific in its coverage of Monkton Coke Works. The issue also had not gone unnoticed by local television news channels, with the public inquiry frequently being discussed on the local evening news programme *Look North*.

However, it was a documentary on the nationally broadcast *Watchdog* that gave Monkton Coke Works a countrywide profile. Reaching a weekly audience of six million viewers, the presenters of *Watchdog*, Lynn Faulds Wood and her husband John Stapleton, would examine problematic experiences that people from around the UK were having with traders, retailers, and other companies. The main cases centred on customer service issues, fraudulent behaviour, or security concerns. The presenters led many investigations that would later succeed in pushing forward changes in company policies and consumer laws. The most notable case was when the show exposed numerous accidents arising from incorrectly wired electrical appliances, which led to a British law being enforced for all electrical manufacturers to supply customers with fitted plugs.[122] Airing in the 8 p.m. prime time slot every Monday, the journalists would look into several complaints that their research team had received from viewers, and allocate a 10-minute portion of the show to an in-depth film focusing on each of the most pressing matters. Each segment would consist of a contextual overview of the issue, interviews with the viewers who had contacted the programme, tests conducted by the research team and, towards the end of the film, a response from the companies involved.

In December 1987, Jennie wrote to *Watchdog*.[123] She knew that gaining national awareness was the next stage of the fight and that a dedicated television segment would put the spotlight on the issue to as broad an audience as possible. She also recognised that only the medium of television could sufficiently convey the proximity of the coke works to the housing estates, the disruption it created in the community, and the suffering it caused the residents. Within a month, Lynn Faulds Wood and her camera crew were in Hebburn to meet Jennie. Following the standard format of the programme, the team carried out its own scientific investigation to determine the level of sulphur dioxide in the area. They also interviewed a wide range of residents to get their perspective on the smoke and fumes that were being emitted daily from the coke works. Residents shared their views on the health problems that they felt the pollution caused and gave details of how the filth was affecting their living conditions.

Community interest in the Hebburn Residents' Action Group had grown to such an extent that it was no longer viable to host all the desired attendees in Jennie's living room. The camera crew filmed Jennie chairing the latest meeting in the function room of a nearby social club. While the cameras rolled, a backdrop of billowing black smoke could be seen out of the club's window. The *Watchdog* team also filmed Jennie doing research into pollution levels at South Shields Central Library, and interviewed her in her home, with the plant just yards away in the background.

A month later, with the nation having just ushered in the New Year, *Watchdog* aired the episode featuring Monkton Coke Works. The film remains a startling exposé on the plight of the residents in the area.[124]

After welcoming the viewers John Stapleton introduces the lead segment for the show, with a still image of the towering facility taking centre stage. He then asks the audience, 'How would you like something like this at the bottom of your garden?' and explains that his co-presenter has been to the North East to investigate more. Cutting to a shot of Jennie walking in a field, the ominous soundtrack of nonlinear strings and foreboding percussion becomes more menacing as the camera pans to the polluting monolith behind her. A voiceover by Jennie states, 'When I came to live here I thought I was coming to live in the country, but look what I got'. We are then presented with a series of syncopated images that display the labyrinthine plant, the overbearing clouds of black smoke they produce, and an arresting close-up of the hellish flares.

The presenter, Lynn Faulds Wood, tells the audience that this is Monkton Coke Works, 24 hours a day, 365 days a year, churning out coke and filling the air with pollution. Lynn then asks, 'So, where would you expect to put such a belching monster? Surely you would have been mad to build a couple of housing estates around it. Yet that's exactly what happened here, and the people who live here hate it.' She discloses that the houses were built 35 years ago and at the time the council told the residents that the coke works would be closing down soon. 'But it didn't, it grew and grew, and people like Jennie Shearan have been fighting the consequences ever since'.

The viewer is brought into Jennie's home. She runs her hand along the windowsill facing the plant and displays five fingers full of black soot. Jennie turns to the camera crew.

> How would you like that? How would you like to breathe that in when you're sleeping? That was dusted four days ago. One day my anger was so high that I phoned across to the manager of Monkton Coke Works. I felt like jumping the fence because I was so angry and he said he doesn't know how it gets into my bedroom. I could soon tell him because it comes from the ovens at Monkton Coke Works. [125]
>
> — Jennie Shearan.

The combination of the images of the coke works, and the close-up of Jennie's soot-covered hand, immediately situates the viewer into the daily reality of life on Monkton Lane Estate. Lynn goes on to explain that the people who live here have no choice because it is a council estate and moving out is not easy.

She also points out with mordancy that while residents cannot burn coal in their own home since the area became Britain's first Smokeless Zone, 'unbelievably the noisy polluting monster on their doorstep is exempted'. A range of residents are interviewed, from a young boy recounting that the fumes come into the school classrooms, to an elderly man who laments that he can taste the sulphur on his tongue.

Once Lynn outlines the history of Monkton Coke Works, and how it was allowed to double in size by the Local Planning Authority, she gives her own opinion on the dirt levels in the area. 'I've never seen muck in my life like the muck around here. Everything I have touched has made my fingers black. But as the people around here say, you can wash your fingers but you can't wash your lungs. What is this doing to their insides?'[126]

Residents are subsequently interviewed to share their thoughts on how the coke works is damaging their health. A young mother with three sons explains that she has started taking tablets for asthma and her children are suffering from chest infections. An elderly gentleman who has to inhale oxygen each day shares his views that the coke works has made his condition worse.

The camera cuts to Jennie heading up a committee in a highly attended function room, as heavy black smoke from the coke works is visible in the window behind them. Barbara is filmed stating that 'I feel that the time has come for the National Smokeless Fuels to put people before profit and this council got off their backside and did something for us'.[127] Jennie adds, 'There are days when I've had to phone Her Majesty's Inspectorate of Pollution, but I've given up. That is a waste of time. All I end up with are large phone bills. No results.'[128]

The film concludes with an interview with a pollution expert who has been commissioned to sample the air at various places around the coke works. Of the pollutants he tests for in his spot check, he is particularly concerned about the concentrations of sulphur dioxide in the area. Levels are shown to be one hundred per cent higher than normal accepted amounts laid out by the World Health Organisation. A chest physician is then interviewed to comment on the health risks of such high sulphur dioxide levels. The doctor concludes that if these levels are representative of a continuing pattern in the area then it would undoubtedly be a cause of worsening respiratory symptoms within the community, more hospital admissions, and possibly an increased mortality rate. Asked if he would like to live in the area, his succinct response is, 'I would not'.[129]

The final word goes to an exasperated Jennie. 'All we get out of Monkton Coke Works is pollution and more pollution and we're just sick of it. We stand for it no more. Either clean it up or close it up. The children in this town deserve a better future than breathing the pollution from Monkton.'[130] In the space of ten minutes, the whole country had now heard about Monkton Coke Works. There could not have been a more impactful and far-reaching way to publicise the issue.

While the *Watchdog* segment represented a turning point in the amount of attention that the cause was getting, it also highlighted a couple of salient issues that the action group were going to need to urgently address. The first was the matter of a disengaged Inspectorate of Pollution, and the second was the lack of documented proof surrounding claims that the plant was damaging residents' health.

The Inspectorate of Pollutions' remit was to monitor plants like Monkton Coke Works on a regular basis and safeguard the health of residents living nearby. A driving factor behind the frustration in Jennie's voice throughout the *Watchdog* segment stemmed from the inspectorate's lack of investigation into the pollution from the facility and the health effects on the surrounding inhabitants. Not only were the regional inspectors' visits infrequent, something even National Smokeless Fuels had noticed,[131] but the timing of their audits would invariably occur during periods of low activity at the coke works.

The inspector's rare trips to Hebburn would correspond with the rare days that clear skies would be free from the appalling fumes that the coke ovens spewed out. The action group were highly mistrustful of this routine coincidence and suspected collusion between the inspectorate and National Smokeless Fuels. On one occasion, a full week passed when no pollution had invaded the estate. Delighted by this unforeseen occurrence, Jennie phoned the inspectorate to thank them for their efforts. The reply that she received confirmed her deepest worries. The official explained that the plant had been monitored for pollution throughout the week by a laboratory firm that was under contract with British Coal and that Jennie should not expect such high standards to be maintained.

Jennie called for a parliamentary ombudsman's inquiry into the Department of the Environment for their failure to supervise pollution from the plant impartially and regularly, but this was to no avail.[132] In a subsequent government review of the inspectorate at the time, a recurring issue that the National Audit Office pointed out was that the inspectorate was significantly understaffed and lacked senior expertise. There were a series of resignations by senior staff around this time, and morale inside the Inspectorate of Pollution was at an all-time low. Chief Inspector Ponsford headed up the team. He was responsible for recruiting staff, regionalising the inspectorate's activities, and unifying the organisation. Given the agency expenditure cuts that Thatcher's neoliberalist government oversaw, Ponsford had been handed a virtually impossible task. Months after he met Jennie personally to explain that short staffing had meant his team could not make regular checks on Monkton Coke Works, he committed suicide.[133] It was a deeply tragic moment in a tragic situation.

Whether it be due to the complicity that the action group surmised due to the lack of unannounced visits, or due to an underlying resource constraint within the inspectorate, it was clear that the inspectorate was not doing enough to sufficiently assess the minacious levels of pollution coming from Monkton Coke Works. The action group resolved to call attention to this issue whenever they had a public forum in which to voice their concerns.

With community support for the cause at its peak, Jennie decided it would be a timely moment to organise a fundraising event to help sustain their fight. The action group put together a garden fête for the residents of Monkton Lane Estate, setting up little stalls for their neighbours to purchase household goods and refreshments, and hosting games such as a tug of war that everyone could participate in. All the residents were encouraged to come in fancy dress. The *Gazette*'s coverage of the day shows Jennie as a pirate, Barbara wearing a *Spitting Image* mask of Margaret Thatcher, and some of the other members of the action group in an array of dazzling outfits.[134] It was a fun day, and such a success that they immediately planned a bigger event that would involve more people in the town.

Throughout Hebburn, there were several pubs that were popular meeting places for the community. Beyond serving alcohol and providing entertainment facilities such as snooker tables and a jukebox, many of the premises also had a connecting function suite. One such pub was The Victoria Park in the New Town. With a stage at the front and plenty of seating, the pub's suite served as a concert room on the weekends, providing evening entertainment in the form of cabaret shows. The local musicians who would headline the events were often imitators of globally famous stars. One entertainer who had a particularly strong following was shipyard-worker-turned-impersonator, Jarrow Elvis. Wearing a white jumpsuit, huge collar, flared cuffs, and oversize belt, he would often sing on the same bill as other performers, who had equally nostalgic stage names that referenced towns in the area. Alongside Pelaw Pitney and Hebburn Cliff, the troupe drew in thousands of nostalgic pub-goers with their 1960s acts. While more appropriately labelled comedy acts than singers, due to their lack of discernible vocal talent, their unconventional renditions and flamboyant dress code made them local celebrities. The act most in demand though was undoubtedly Jarrow Elvis, with his self-confidence and dress sense more than making up for his singing ability.

The Hebburn Residents' Action Group organised an evening of entertainment for Hebburn at The Victoria Park, and Jennie persuaded the venue to waive their fee and Jarrow Elvis to perform for free. Beyond the proceeds from the ticket sales, the action group also raised money from a raffle that was held before 'The King' hit the stage. It was another lively event that brought the community together and raised much-needed funds for the action group.

On this occasion, both a film crew and a professional photographer were in attendance to record the evening as part of a set of programmes that were in development regarding the issue. The *Watchdog* segment from earlier in the year had aroused interest in Monkton Coke Works throughout the North East.[135] Tyne Tees Television, the local arm of ITV, had commissioned an in-depth 30-minute documentary on the matter, to cover the cause in more detail on their *First Edition* series.[136] Furthermore, a talented photographer, Keith Pattinson, who had captured engrossing images of the Miners' Strike earlier in the decade, had been tasked with bringing this subject to life with a series of photographs for an upcoming art exhibition at the nearby Side Gallery.[137]

The issue that had blighted Monkton Lane Estate for decades, was now a firmly established fixture in the media. The Hebburn Residents' Action Group had instilled faith in their community that their voices could be heard and now Monkton Coke Works was a regular focal point of discussion in both the local press and local news shows. Jenny Johnson made front-page news when the *Gazette* interviewed her to get her views on how life had been adversely affected.

> The smell is enough to make you sick. The coke works are burning our lives away. I rang Alan Ridley of Her Majesty's Inspectorate of Pollution and told him that

my fish pond had to have its water changed daily and that the fish were barely visible through the murky grime-filled water. He replied, 'You shouldn't have a fish pond if you live near a coke works'. A lot of people around here need oxygen masks to help them breathe and they were never smokers either.[138]

— Jenny Johnson.

One newspaper led with a headline that likened life near Monkton Coke Works to the recent Chernobyl disaster.[139] Another publication ran a visual feature on the plant while it flared off excess gas, explaining that the site had become a beacon for motorway drivers approaching the region, who used it as a signal that they were now in South Tyneside.[140] A resident wrote in to the *Gazette* with a heartfelt poem expressing his anger at Monkton Coke Works.

Grimy embodiments of tallness,

Soot galvanised chimneys.

Silhouetted against a cancerous haze,

Belch out putrid and obnoxious clouds,

Like flatulent demons.

With the same manners,

They nonchalantly refuse to excuse themselves.

Inexorable tongues of flame lick

The skylines' boots like con-men,

Exchanging sulphur dioxide for oxygen,

While confused, surrounding fields live

In perpetually enforced daylight,

Their silent expostulations rewarded

With a topsoil of black faeces.

The ageing and coughing worm turns,

As Monkton Coke Works,

Hebburn pariah,

Attracts protests.[141]

By now there was no way that British Coal or the Inspectorate of Pollution were unaware of the chagrin that the residents felt towards the coke works. When the 30-minute Tyne Tees Television documentary was subsequently aired, it was a neat summary of the status of the issue.[142] The title of the show was *Bad Neighbours.* The term 'bad neighbour' had become embedded into environmental legal language. This expression referred to a facility which belonged to an essential industry that had strategic benefit to the country and provided employment, but also damaged the environment and harmed public health. Monkton Coke Works ticked all the boxes.

The opening words were given to Jennie, at her home, just as the *Watchdog* segment had done. As she spoke, the view from Jennie's bedroom window displayed the conspicuous coke works.

On top of the noise and smell, we have the ongoing saga of washing.

As you can see, I've just taken this net curtain down. It's only been up a matter of eight to nine days. You wash it constantly. After a few washes, it gets discoloured and you have to buy new. You also have your paintwork which is just filthy.

This is why we formed an action group. Not only does it come into our rooms, but this is also the bedroom where I sleep. So if it goes onto my windowsill, am I breathing it in? What does it cause when I breathe it in? It must have a health effect on people living in this area. It never lets up.[143]

— Jennie Shearan.

Once again, the plant itself featured prominently throughout the film, with the overwhelming sirens and the inordinate quantities of dirty smoke assaulting the screen. The dichotomy of the nearby housing, only a few hundred metres from the site, was accentuated throughout the documentary by interviews with residents. Stories of chest complaints, sleepless nights, and tightly locked windows were shared.

There were also light-hearted moments during the programme, with footage from the fundraising evening that the action group had organised. In between clips of Jarrow Elvis serenading audience members, whose reactions ran the gamut of being bemused to captivated, attendees were asked to share their thoughts on the cause. Tellingly, one older gentleman spoke very highly of Jennie, saying, 'No one had the guts to do what Jennie Shearan has done'.[144] The closing words went to the action group's leader, who summarised their stance in her typically evocative style.

> Over the years we have watched it grow from something very small to new buildings all the time. We have had enough.
>
> This is the people in Hebburn's fight. It's not my fight. I'm just a person that is trying to get people to come together.
>
> I have been fighting for something which I think is everybody's God-given right, which is to be able to breathe clean air. There are thousands of people in South Tyneside unemployed. I don't want to add to this list. All we want is to clean Monkton Coke Works up because it can be done. If you can send a man to the moon, you can clean up Monkton Coke Works. And it all boils down to spending money to do it.[145]

— Jennie Shearan.

Autumn arrived, and it was clear that 1988 had been a breakthrough year for the Hebburn Residents' Action Group. Their anti-pollution campaign had moved from coverage in the local press to national television. Awareness of the issue had been amplified throughout the region and community support for the activists was substantial. Still, it had come at a price for the action group, and especially for Jennie.

In life, whatever we choose to do comes at the expense of not being able to do other things. It is the choices that we make that define who we are, and Jennie had felt compelled to prioritise this cause for so long because she believed so strongly that her community deserved to breathe clean air. The path that she had chosen had forced her to singularly focus on this one ideal, to the unavoidable detriment of other areas in her life. She was incurring costs to her health, her wealth, and her family's well-being.

Jennie had developed angina, which is a type of chest pain caused by reduced blood flow to the heart.[146] It is triggered by stress and high blood pressure. Her monthly phone bills were hundreds of pounds due to frequent calls in pursuit of her goal, and the endless photocopying, posting, and photo-developing had eaten into her savings. Jennie's ailing husband David was a constant source of encouragement, but he had become increasingly unwell and she was struggling to balance taking care of him and pushing on with the campaigning. She was 66 and exhausted. One journalist quoted her from the time, with Jennie saying, 'My doctor told me to take it easy. But the problem is that if I don't continue, who will?'[147] Fearing that a generation of residents was being doomed to live a life of ill health, Jennie's determination knew no bounds. If there was a particularly invasive black push, she would spend untold hours recording the whole event from multiple angles, using her camcorder. If it was past midnight and Monkton Coke Works had caused a bad polluting incident, she would put a coat over her nightclothes and visit the site manager to get the problem fixed. On one occasion, Jennie and a neighbour visited the plant at 2.30 a.m. to complain about the noise and smell that had kept many in the community awake that particular night. When the police were called by Monkton Coke Works to remove the protesting trespassers, Jennie told the policemen,

'If I'm trespassing then so is the company, because the noise, filth, and smell from the coke works are coming right into our homes'.[148] The police officer agreed with Jennie and drove off.

Despite the widespread support from her community, her fight was not universally backed by local neighbourhoods. As the very public face of the issue, and at the forefront of the battle for a healthy environment, Jennie encountered tension and resistance from some sections of the town who did not fully embrace her efforts. She could shrug off the sneering remarks of some men who attempted to diminish the action group's activism as merely a series of complaints about dirty laundry. However, she also had to experience negativity from residents who did not approve of her fight; there was a small portion of the community who equated Jennie's campaign with a risk to the livelihoods of the men who worked at Monkton Coke Works. The Miners' Strike was still fresh in many people's memories. The loss of income that those on strike had to accept for nearly a year was a scar that had yet to heal for many, including the men working at the coke works. For some of the 200 workers who had returned to employment at Monkton Coke Works, Jennie was not a popular figure. Even though Jennie just wanted clean air for the community and did not want the men to lose their jobs, some of the workforce believed she was trying to close down their employer and take away their jobs. She was not fighting the men; she was fighting what they were employed to produce. Nevertheless, it felt like a dismal zero-sum game in which someone in Hebburn was destined to lose.

These were difficult circumstances for Jennie, but she felt she had no option but to continue making sacrifices. In October 1988, her dear David, who had been suffering from pneumoconiosis and another stroke, passed away from heart failure. It was a devastating loss for Jennie. They had been happily married for over 40 years, built a family together, and shared many beautiful memories. With her children having long since flown the nest, and her husband no longer with her, she was, for the first time in her life, living on her own.

CHAPTER SEVEN

ON THE ROAD

**'These houses shouldn't have been built so near to
the coke works.'**

The months of mourning after losing her husband were heart-wrenching for Jennie. Although her children were there to support her throughout her bereavement, it was a profound loss for her and the ebbs and flows of grief would continue for many years. There were many days when she was consumed with sorrow. Slowly, with the help of her children and friends, she picked herself back up.

With so much to still fight for in her battle for clean air, Jennie knew she had to continue with the anti-pollution campaign. There was a nagging issue that had perturbed Jennie and the action group in the immediate aftermath of the public inquiry, and the months that followed. The news during the public inquiry that the European Commission had granted National Smokeless

Fuels significant funding for their proposed power generation station did not make sense to them. Jennie could not understand why the European Community would support a scheme that in her mind would exacerbate pollution in her area. Indignant that the governing body had not seen the bigger picture, she set out to learn more about the organisation that had authorised the decision.

Established in 1958, the European Commission is an executive branch of the European Union, divided into departments with specific zones of responsibility covering everything from economic and financial affairs to energy and the environment. The European Union (EU), was originally the European Economic Community (EEC), which had grown out of the European Coal and Steel Community. Founded in 1957 to drive postwar prosperity and peace among its member states, the EEC was birthed as a multinational union that principally aimed to bring about economic integration among its country members. Significantly, its purpose extended beyond free trade. One of its foundational objectives was to improve the quality of life of the inhabitants of its member states, and that included protecting their environment.

The EEC originally consisted of six European countries, with the UK joining later, in 1975. Under the founding treaty which established the EEC, various instruments were created from which laws were made within the EEC institutions. To have a level playing field across the countries that had differing levels of economic power, directives were drawn up in Brussels, the de facto capital of the EEC. In essence, directives were an instruction to the governments of member states to pass various obligations into law in their domestic legal systems. The intention was that citizens of member states would be able to access legal rights and obligations once the relevant domestic laws had been passed in accordance with the requirements of European law.

There were three political institutions that held executive and legislative power across the EEC. The European Council represented the state governments, the European Parliament represented citizens, and the European Commission represented European interests. The European Council or European Parliament would place a request for legislation to the European Commission. The European Commission then drafted the laws and presented them to the members of the European Council for approval and members of the European Parliament for an opinion. From there, the laws would be signed off and transposed into the laws of the member countries.

The European Court of Justice was then set up to interpret these laws and ensure its uniform application across all member states. In the early years of the EEC, the European Court of Justice was very active in promoting a system of rights and obligations to remedy the deficiencies of European law where member states had failed to properly implement it. One such doctrine was 'direct effect', where citizens could seek to enforce rights granted to them under European law. This was a revolutionary concept because rights made within international organisations were generally not granted to individual citizens through their courts.

The European Parliament is composed of several hundred democratically elected members from the participating countries. When the UK was still a member of the EEC, it had representation for each of its sub-regional constituencies, with Joyce Quin the representative for Tyne and Wear. Every month, Quin, along with the other members, would meet at the Parliament headquarters in Strasbourg for a four-day plenary session. The plenaries were the best place for members of the public to be able to meet with members of the European Parliament and officials. The other weeks of the month were spent in Brussels, in committee meetings. Through Jennie's unceasing correspondence with politicians, journalists, and dignitaries, she had developed a relationship with Quin, and Quin had acquired a strong understanding of Jennie's anti-pollution campaign.

In discussing the Monkton Coke Works issue in depth with Quin and her colleagues, Jennie had learned that prior to their planning application, National Smokeless Fuels had successfully applied for an EEC funding pot that was designed to support economic development. Having been presented a case for the economic benefits of the power generation station, the European Commission's Energy Committee had deemed the firm eligible for the European Regional Development Fund. The grant was intended to be split between the upfront capital and ongoing operational expenditures that the power generator initiative would incur.[149]

Armed with this information, Jennie felt it imperative to enlighten those same decision makers about the residents' side of the story. As much as National Smokeless Fuels were unwavering in their intent to expand the plant, Jennie was unstoppable in her quest to secure clean air. She resolved to take the fight to the European Parliament. Jennie's presumption was that if there was a grant to support the creation of a power generator, then there could also be a grant to finance a clean-up of the coke works. At the very least, Jennie wanted the Energy Committee to review pollution levels from the plant to check whether emissions met European environmental standards before any further grants were offered to National Smokeless Fuels. Quin recognised the importance of the European Parliament having a balanced view on the matter, and to investigate the health issues stemming from the facility. She helped broker a series of meetings between the Hebburn Residents' Action Group and Members of European Parliament and key European Commission officials, to take place in February 1989. Jennie was going to meet the very people who were responsible for framing future environmental legislation.

This was an exciting development for the Hebburn Residents' Action Group, who were very appreciative that they would soon be able to get their voice heard at such an elevated level. With the visit to Strasbourg only two months away, the race was now on to raise funds to enable a delegation to travel to France. Jennie turned to South Tyneside Council for financial support. The action group had borne all the costs over years of campaigning and needed assistance for the travel expenses they were about to incur. Unfortunately, no aid was forthcoming from the council, who deemed this a political issue that they did not want to be involved in. Jennie's dismay for the council's reluctance to support the residents on a health matter of such vital importance to Hebburn was summarised in her pithy pronouncement to the *Gazette*: 'It's not political looking after your health and the health of future generations. Someone's got to do something and we won't rest until it's been done.'[150]

The lack of funds was just another bump in the road that would not deter Jennie. She had become so accustomed to facing hurdles that her default response was to simply explore other avenues in this latest step of the campaign. The prominent status of the Monkton Coke Works issue meant that many people throughout Tyneside and across the North East were aware of the action group's cause, including well-known personalities who had become successful leaders in their field. Jennie felt that these people could be a possible source of help for the activists.

One such successful local figure was Catherine Cookson. Much of Cookson's work had been adapted for television, radio, and the stage. The author remained devoted to her birthplace throughout her life, and generously pledged large sums of money to a variety of causes to help uplift the North East. Her philanthropy reached medical, cultural, and educational institutions across the region. As a key matter that had featured regularly in the newspapers across her region, Cookson had followed the Monkton Coke Works problem.[151]

On receiving Jennie's request for support, Cookson's response was immediate and a £500 cheque,[152] worth around £1,200 in today's money, was delivered to the Hebburn Residents' Action Group. With a cover note sending Jennie best wishes, the letter also stated, 'I did try to phone you, but it seemed to be continuously engaged'.[153] Jennie's phone would have undoubtedly been occupied with a wide range of phone calls to secure support for the trip.

Over the years of campaigning, Jennie's black book of media contacts had mushroomed from a small local network to a range of influential supporters throughout the country. The national broadsheet *The Sunday Times* had picked up on the story about the action group's upcoming visit to the European Parliament and launched an appeal to raise funds for 'their acrimonious fight against the dark satanic mills'.[154] Donations, however small, were gratefully accepted, with further monies being generated at a collection at a nearby pub.[155]

The action group now had enough charitable donations to send a three-person deputation to Strasbourg for a week of meetings.[156] They decided that Jennie, Barbara, and Jenny Johnson would be the triumvirate to make the journey. In the weeks leading up to the trip, they started gathering all their evidence into a package that they could present succinctly to the Members of European Parliament. Press cuttings were assembled and chronologically placed into a scrapbook. Sound recordings of the noise pollution were aggregated onto one cassette. Jennie dug up fresh soil from the playing fields surrounding the coke works so they could be analysed by scientists. The action group raised a new petition, gathering 1,200 signatures in support of their clean-up campaign. A dossier was compiled using the stirring photos that Keith Pattinson had taken to help tell their story. The report was entitled 'Hebburn Residents' Action Group – Campaign Against Pollution by Jennie Shearan, Chairman', and led with an invitation for the EEC to visit Monkton Lane Estate. Within its fourteen professionally designed pages, commentary on the background of their cause was blended with anecdotes from residents to provide a comprehensive overview of their predicament. A few days before the trip, they even organised a photo opportunity with the South Tyneside Courier to pose with a copy of a recent article that the newspaper had printed about their cause, and brought along the subsequent front-page headline screaming, 'FIGHT GOES ON'.[157] Aged 67, Jennie was at the peak of her activist powers. The Hebburn Residents' Action Group was ready to lobby the European Parliament.

Strasbourg is placed at the very heart of Europe and is a fitting location to discuss matters pertaining to the EEC. Situated at the eastern border of France with Germany, its 200,000 inhabitants are just a 90-minute drive from Switzerland to the south, or Luxembourg to the north. A flight from the action group's nearest airport to Paris would have taken 90 minutes, followed by a further 90-minute train ride to Strasbourg. Alas, the protesters had to conserve their funds.[158] Their pilgrimage lasted over sixteen hours. Starting with an early bus from Hebburn to Newcastle, the main city of the North East, they then took a nine-hour coach down to Dover, followed by a two-hour boat ride to Calais. Once in France, a further five-hour train ride got them to Strasbourg. After a full day of travelling, they checked into their budget hotel where they shared one bedroom between the three of them.

The journey may have been arduous for the trio, but their energy levels remained high. As they approached the fifteen-acre site where the Palais de l'Europe stood, they were reminded just how far they had come, literally and metaphorically. Before entering the fortress-like building, Jenny Johnson took a photo of Jennie and Barbara. Standing in front of the tall flag poles that adorned the front lawn, the photo would serve as a visual reminder to the mother and daughter of the extent they were willing to fight for their community.

The impressive building did not intimidate them; it invigorated them. The aluminium-covered façade, the vast windows, and the iconic hemicycle represented hope for the action group. Their overall remit was clear: gain support for the environmental clean-up of Hebburn. Jennie was convinced that it was National Smokeless Fuels' unwillingness to invest that was preventing the reduction of pollution in the area. She knew that ultimately, the cause and resolution of this issue came down to money. As such, the action group had four distinct funding requests for their European representatives. They wanted aid to pay for double-glazed windows to be installed in the nearby houses to help block out the intrusive noise that was particularly thunderous at night. They wanted compensation for residents around the plant whose health problems had been aggravated by pollution.[159] They also wanted any future monies that would be provided to Monkton Coke Works to contribute towards environmental improvements at the facility. Finally, they wanted funding for a health study to be made of people living near the plant.

Since gathering the original petition of residents after the first public meeting with National Smokeless Fuels, the action group had harried South Tyneside Council, demanding a full-scale health probe. The ever-increasing pressure from the action group had been noted by the council but no progress had been made. While support for a health survey was arguably the most important request of the European officials, the action group also took the opportunity to state the lack of monitoring at the plant, in an effort to persuade the EEC to bring pressure to bear on the Inspectorate of Pollution.

Over the course of five days, they met with several commissioners and parliamentary members, stopping off at the Brussels headquarters for further meetings before embarking on the long journey home. The consensus from the European Commission chiefs was that the campaigners' case warranted deeper investigation. Jennie had put Monkton Coke Works on the European radar.[160] On her return, the local press were curious to get Jennie's take on the trip.

> I am annoyed at South Tyneside Council [for] not giving us a grant to go to the European Parliament as we are fighting for local people and we are trying to get more money for South Tyneside.
>
> The council has done nothing about the coke works and they always refer us to the inspectorate, who in turn say it is nothing to do with them and refer us back to the council.
>
> I think everyone in the European Parliament now knows about the Hebburn residents' fight to be able to breathe clean air.
>
> We put our points across to many members of the European Parliament and spokesmen for the Environment and Energy Committees.[161]
>
> — Jennie Shearan.

Jennie knew that the visit and her subsequent comments would be embarrassing to South Tyneside Council, but she did not care. She herself had been Hebburn's councillor for many years. She knew what the role meant to the community, and what it could achieve with hard work. Jennie felt let down by the very people who were meant to represent her constituency.

In the weeks following the action group's trip, some heartening news emerged. South Tyneside Council commissioned an official health survey of families living on Monkton Lane and Luke's Lane Estates.[162] People power had once again forced immutable institutions to change their long-held policies in favour of the action group. This would be a major two-year inquiry into the health of people living close to Monkton Coke Works. At the behest of South Tyneside Council's Environmental Health committee, the borough's health authority was given the mandate to study the health records of hundreds of families. Jennie was confident that the results would provide ammunition to secure a complete clean-up of the plant. A significant budget of £50,000 was set aside, £120,000 in today's money, and the borough's leading physician was deployed to oversee the project.

Alarmingly, in the weeks leading up to the survey's scheduled May 1988 commencement, a debate emerged between the governing bodies who had a vested interest in the future of Monkton Coke Works. The Inspectorate of Pollution stated it could not back the scheme since it did not recognise a serious health hazard from the plant. In a revealing indicator of where their priorities lay, they also pointed out a concerning loophole in the legislation that did not consider emissions from coke works to be harmful or offensive. In alliance with the inspectorate, a representative from the National Union of Miners, who represented the workers at Monkton Coke Works, declared that he had not heard of health problems from workers in the plant. The self-serving motivations of the powerful inspectorate and union were clear: dismiss any health concerns and keep the plant in operation. Coincidentally, on further investigation into the anticipated cost of the survey, the projected figure that was discussed had now doubled to £100,000.[163] Just as soon as the health survey had been commissioned, it was deemed too expensive, and thereafter cancelled.[164]

This was a major setback, and Jennie was dismayed. 'They should come and talk to the residents with inhalers and oxygen masks under their beds.'[165] Without delay, the action group reacted in the only way they knew how. They decided to conduct their own survey of the whole of Monkton Lane Estate and its surrounding streets to gather medical evidence on the repercussions to the residents' health. They were certain that the coke works had inflicted a wide range of detrimental effects on the health of local residents and they wanted to verify it to understand the extent of the damage, with or without the support of the council.[166]

The action group created a typewritten questionnaire on A4 paper, entitled 'Campaign against pollution'. Their questions were thoughtfully and succinctly worded, and the layout was simple, with ample space for answers. After the section to input name and address, six questions were listed. 'How many people live in your home?' and 'How long have you lived here?' were followed by two multiple-choice questions: 'Do you think your health and well-being have been affected by living close to the coke works?' and 'Do you think that since 1981 the smoke, noise, and dust from the coke works has reduced, increased, or stayed the same?' The fifth question asked the participant to tick from a range of illnesses from which they were suffering and any medical devices that they may be using. The final question asked if the residents believed that these health problems were caused or aggravated by Monkton Coke Works, or that the plant made no difference. At the bottom of the questionnaire, there was a section for any other comments.

The Hebburn Residents' Action Group knew that their heuristic approach had flaws. They were not scientists and had never conducted a health study of this kind before. They appreciated that establishing a correlation between the facility and the residents' worsening health would require medical professionals to carry out a thorough investigation and take into consideration many variables. However, this was not an option for the group. They needed to capture the convergent opinions of every resident who was living in the vicinity of Monkton Coke Works. The group set about delivering the makeshift survey by hand to the 1,600 houses in the Monkton Lane Estate area and the other properties within a two-mile radius.[167] Each leaflet contained a cover letter explaining the need to collect information on the health and well-being of the residents, given the lack of support from the local council. The note stated that the questionnaire would be collected in three days' time and ended with a sign-off from Jennie as the Chairman of the Hebburn Residents' Action Group.[168]

The printing and posting of the forms was a big undertaking. The high amount of paper and ink required was costly. Moreover, it took the action group over 60 hours to hand deliver all the forms to each household. Encouragingly, the response was immediate and positive. Enthusiasm built across the housing estates as neighbours grew aware of the action group's initiative. Some residents who had not yet received the form began asking Jennie for a copy of the questionnaire. Others donated money to the action group to help them with the costs of printing such large volumes. The understandably impatient residents were assured that they would receive one within a day or two, and the benevolent donors were thanked for their kindness. The action group were being as methodical and thorough as they could, within the limited means that they had available to them.

> My health is not so good these days and my bank balance is rapidly approaching zero, but we must do everything we can to stop this pollution of our homes.
>
> I didn't believe the situation could get worse, but it is, every day.
>
> When people get the forms, I plead with them to fill them in honestly to give us a chance to have the results properly analysed which I am convinced will show the justification for a full-scale medical survey.[169]
>
> — Jennie Shearan.

Over the following days, the majority of the forms were collected as complete, representing over 2,000 adults, and 600 children. The Hebburn Residents' Action Group now had evidence from a survey population of thousands of residents about the reality of living next to Monkton Coke Works.[170] While they analysed the feedback, Jennie expressed her scorn towards the council to the *Gazette* journalist who by now had a hotline to the action group's leader.

We are doing the work they should be doing. I'm very sad because the council didn't think that this survey was important enough for them to take it on board.

Our group feels that people want the survey and it is worthwhile.

We are amateurs but we are going to manage it. Our budget is very low in terms of cash and high in terms of time. So far, ninety per cent of our questionnaires have been returned as completed.[171]

— Jennie Shearan.

When the results were fully collected and analysed, they told a stark story. Seventy-eight per cent of residents felt their health and well-being had been affected by living close to the coke works. Eighty-four per cent thought that since 1981 the smoke, noise, and dust from the plant had increased.[172] Over 1,000 residents believed that Monkton Coke Works had caused or aggravated their health problems.[173]

This was hard-hitting data in itself, but it was the hundreds of written comments from residents who had lived in the area for decades that painted a tormented picture of the all-pervading nature of the coke works. Often the residents did not just fill the three lines that were given to them in request of any further details, but also filled the blank page on the reverse of the form as well, expressing a multitude of detrimental effects on their physical, mental, and emotional health. It was an outpouring of stories from residents who had been abused by the inhumane presence of a coking plant that had a round-the-clock stranglehold on the community. The following is a selection of responses from the survey.

> Now I know why the man told us Hexham Avenue was called Death Row. I remember that day. He says to me, 'Do you know what they call this Avenue?' I say, 'Yes, Hexham Avenue.' He says, 'No, they call this Death Row.' And I thought he was joking.

> Apart from health problems, the filth emitted from the coke works is a constant headache. So much for the Smokeless Zone! It is beyond credibility that a council should build family houses in such close proximity to the coke works. It is ironic and stupid, that while ordinary households have to comply with the regulations, the coke ovens are still allowed to belch forth tonnes of filth.

> The smell that comes off the gas is just intolerable to bear. We keep windows closed when they should be open to give us clean air to breathe. North East is left out as usual. If it had been south of the country there would have been no problems having it seen to. Everywhere around the house is always black with the dust. It costs us a lot of money for decorating all the time and changing nets and curtains as they are never clean for more than a couple of days and if this goes on items such as these, what are we inhaling?

The benefit of the Miner's Strike was the clearer air. It was just as though we had been given a new lease of life. You could leave your clothes out and your windows open. Since then, the coke works have gone berserk. There is double, maybe treble, the filth and pollution, with the noise, smell, fumes, smoke, and dust.

It is difficult to imagine that the thick clouds of smoke and nauseous stench given off by these works are not doing any harm to the people in the area. You can be sure of my family's support in your campaign! Sometimes I can't breathe properly because of the smell.

My health has deteriorated since moving onto the estate. The coke works cause me to sneeze and cause soreness to my eyes. Every day I wake up with a sore throat and a cough and my chest feels tight. The smell at times is terrible, you would just think a chemical gun had exploded.

Filter the filth from the ovens! If the plant cannot be modified to an acceptable standard, then I believe it should be closed down. It should be re-established in an area that is not so heavily populated. I am worried about the long-term effect on my children.

I have a new baby who I cannot leave outside because the dust gets in his creases and makes him sore. All my washing is hard after being on the line. My baby's wool clothes are ingrained with dust and I can't let him play on the lawn because of the coal dust. Have you ever tried getting to sleep when they're pushing the ovens?

We are very concerned both with the increased amount of sulphur and other gases which are being constantly emitted into the air. Although we do not live in Monkton Lane Estate, the prevailing winds mean that the pollution is often directed into our estate. When the wind is in a certain direction, the smell from the coke works is nauseating.

The pollution is certainly intensified at night time whilst people are asleep and dust is everywhere if the windows are left open. The pollution also definitely increases during the weekend.

The thick black smoke that is released from the coke works only used to be released during the night when the people weren't supposed to see it. Now it is released during the day. Sometimes it is so bad it could cause accidents on the main road. My car, windows and gardens are all recipients of the coke works free gift: filthy black deposits.

You'll have about four big jets of yellow gas, all the sulphur and everything, and one day I went out in the garden and I didn't realise, and I got a mouthful and believe me it did me no good. I was really bad with it, and now I watch out, even if it's just hanging something on the line. I don't go out the back much at all.

The main problem in this area is noise during the night shift. It's a bloody nuisance. Also, the smell is sometimes unbearable. It stinks.

On windy days, I find that plants in the garden seem to have a fine dust on them and some years no plants grow at all and just die off, even after a good show the previous year.

We lived in Hexham Avenue for eight years until we moved after each one of us suffered in some way, so we were forced to move for our children's sake. The smell and fumes and muck were disgusting. We feel something should be done for the people that are left with what we have escaped from.

Since moving into the area I am alarmed at the amount of thick black dust that accumulates, possibly in your lungs. If the weather's heavy I have a job breathing. When it hangs from the coke ovens it's terrible, it seems like you're chewing on it. We have double glazing and the coal dust gets in between, even when we have the window closed. So we must be breathing that in, night and day. I have even woken up in the middle of the night with the smell of sulphur.

The smell from the coke works is disgusting. Even after washing, the clothes still stink. The smell gets in the house. That can't be good for anybody's health. If the wind blows from Monkton Coke Works my children are kept in and my window closed. I live a mile away. God help the people that live on the street near to it. The best thing that could happen is to get rid of that filthy place. People need to breathe fresh air again and smell nature again and not the coke works.

Why should I suffer another 25 years? Why should I suffer another 25 minutes? We are living in a Smokeless Zone and have to suffer the disgustingly foul smell and black soot which blows our way from the coke works. This cannot be conducive to a healthy environment.

There's a fine layer of soot in all rooms in the home. It is even in every bed. The amount of washing, mostly whites, that are ruined by black dust is so annoying. Also, paintwork gets covered in it. If it does that to paintwork and clothes, I hate to think what it does to mine and my kids' insides.[174]

Years of disruption, isolation, and humiliation were now articulated in heartrending black and white. The pervasive effects of four decades of airborne pollution raining down on a community had been set down on paper. The Hebburn Residents' Action Group had given their people a chance to express themselves, and the response was visceral. Their basic human right to clean air had been violated and they finally all had a channel through which each could communicate their pain. The feelings and stories that were compressed within the lines of the surveys stated every facet of Monkton Coke Works' destructive assault on the area. The plant had dominated their lives, through its visual, olfactive, and aural impacts, and the deleterious effects of its pollution on their health. The residents had carried the burden of the environmental costs of coking for decades, and finally had an outlet to tell their stories.

The responses showed that residents felt scared by what the pollution was doing to their health, with a staggering number of bronchial and sinus problems reported.[175] They also revealed the daily reality of feeling trapped inside their homes, with doors and windows permanently closed to minimise exposure to the air laden with fumes and grit. The comments disclosed how the residents felt discriminated against by the power disparity of a perceived North-South divide, and the inequitable distribution of environmental risks and economic benefits. Finally, their feedback indicated how they felt neglected by a system that had given them no means to redress their situation. The towering pile of completed forms were a clear and tangible evocation of the total discordance of placing a housing estate next to a coke works.

The action group promptly drew up a summary of the key statistics and the recurring resident feedback. Their overview told a powerful story and in Jennie's words, 'made for very sorry reading. I feel that for 180 jobs the price we are paying is too high, and I and the other members of the action group are not prepared to pay it any longer'.[176] The message from the action group had clearly evolved. Whereas they had originally solely focused on the need for the coke works to be cleaned up, they began punctuating their efforts with a call to close the plant if it could not eliminate the pollution. They did not want to put men out of work, but they felt that given the choice between health or jobs, health should come first.

The action group felt compelled to share their stance and the insights gleaned from the health survey with key influencers who had a platform of their own from which they may be able to voice support.[177] They wrote to Thatcher, asking for help. Neil Kinnock, Labour's Leader of the Opposition also received a request. Dozens of ministers received packages from the campaigners containing the results. They also contacted charitable organisations that strived for a greener planet, such as Greenpeace and Friends of the Earth.[178]

Throughout her time as Chairman of Tyne and Wear County Council, Jennie had met several members of the Royal Family, and the action group had written to the Queen to update her on the campaign's progress. One key figure in the Royal Family who Jennie had had the chance to meet on several occasions was Prince Charles. The first in line to the throne was a passionate champion of environmental issues, and in their brief exchanges, Jennie had detailed the problems that the residents faced on a daily basis. On one occasion Prince Charles had told her, 'Never give up fighting for the community and never change. Always be as caring as you are'.[179] The Prince duly received a summary of the action group's health survey results, along with a dossier of recent press cuttings.

The action group also shared these insights with their key journalist contacts. Every single local newspaper printed at least one article on the evidence that the action group had accumulated, with Jennie making herself available to answer any follow-up questions or share her views on the results. Scathing headlines such as 'Doorstep survey reveals coke peril', 'Coke works pollution causes health problems – survey' and 'Works are unhealthy says survey' dominated the leading pages of the *Gazette*, the *Sunderland Echo*, and *The Journal*.

The published findings also heralded a step-change in the frequency with which stories were being published about Monkton Coke Works and the action group's campaign. On an almost weekly basis, there were reports across the regions' press that featured the issue. In one article that outlined residents' lives as being 'plagued by the pollution and smells', Jennie forthrightly stated, 'The basic problem is that houses should never have been built so close to the works in the first place'.[180]

In another article, the journalist explained to the reader that the previous evening, a strong outbreak of black smoke had emerged from Monkton Coke Works. Clouds of dust filled the air around Monkton Lane Estate, and Jennie, Barbara, Jenny Johnson, Marian Carey, and Maudy Cork decided urgent action needed to be taken. The five-strong group marched to the site and refused to move until a senior member arrived.[181] They waited for three hours at the plant, sitting patiently in the staff room for a plant official to arrive. While ultimately no management arrived, it did provide for a good story and the following morning, the *Gazette* published the story. The article led with 'Angry women stage sit-in', accompanied by a photo of the protesters defiantly staring into the camera lens.[182]

In a separate piece focused on Jennie's efforts to lead the campaign, her comments summarised the progress that the protesters were steadily making.

> It's taken a lot out of me recently. But we are begin-
> ning to get results. We can achieve our aim of either
> having Monkton Coke Works cleaned up or closing
> it. People are more important than profit, and a clean
> environment is certainly worth fighting for.[183]
>
> — Jennie Shearan.

Running parallel to the editorials focused on Jennie and the action group, multitudinous articles began to get printed that explored other residents' perspectives. It was as if the topic had acquired a life of its own. The snowball that Jennie and her team had laboriously pushed up the mountain for years had now emerged on the other side to create an avalanche of press coverage.

Terry Kelly, the young *Gazette* reporter who so diligently followed Jennie's every campaign move, led the way. With con-demnatory headlines such as 'Hell-hole',[184] his reports ranged from a two-week-old baby who coughed up black coal dust[185] to news that a nearby school was going to be fitted out with a volumetric gauge to measure sulphur dioxide levels.[186]

Residents began coming forward to the press with their own stories. John Errington furiously described how the pollution had become more and more unbearable.[187] Jack Douglass likened the fetid smell in the area to a 'bad case of the morning after a booze-up on Brown Ale'.[188] Amy McCarthy, who relied on an oxygen mask to breathe, recounted how she had to drive out of the area one weekend in search of fresh air.[189] Elsie Turner opined that 'the directors of the plant should have their large executive homes and their children's schools and playgrounds built where Jennie Shearan lives'. John Badger deduced that 'the courageous stand of Jennie Shearan's little band' showed that 'we can no longer have any confidence in our elected representative in these matters: their priorities are clearly not ours'.[190] Jack Wandlese wrote,

> The coke works produce more pollution now than ever before. All the other works closed down throughout the country and now have their production done at Monkton. Mrs Shearan and her colleagues are fighting on behalf of us all, they want the pollution cleaned up, not the works closed down. I must point out that I am in no way connected to Mrs Shearan or her group but I do admire what she is doing on my behalf, and good luck to them.[191]

Margaret Landless pointed out that the workers at the plant now wore protective clothing, masks, and earplugs, but the residents had no such defence against the pollution.[192] Raymond Bay questioned what safeguards were being put in place to prevent a health tragedy at Monkton Lane Estate. Thelma Young wrote a letter to the editor expressing her sorrow for the residents who had to contend with a health hazard that produced such awful grime and dust. Hayley Snaith asked why 200 jobs were being valued higher than thousands of people's health.

Jennie had empowered the community to stand up and be heard. Reputable names in the green environment movement also joined the chorus of support. Jonathon Porritt was a national figure who, in his role as director of Friends of the Earth, had multiplied the charity's membership tenfold. He had closely followed the progress of the action group and commended the 'classic model of the need to act and address environmental problems'[193] that they were pursuing. Similarly, the environmentalist segment of the region's leading newspaper often highlighted the action group's plight. Branded 'Greenwatch', the recurring editorial frequently chronicled the anti-pollution activists' campaign against the coke works that loomed large over the estate.[194]

Despite the self-propelling velocity of coverage that the matter was now generating, Jennie knew that this was not a time to wind down her own efforts.[195] With her supporters taking on the mantle as outspoken standard bearers, Jennie began making regular visits to the Inspectorate of Pollution's headquarters in London. They were long days. To save money, she would always make her visits to the capital a day trip. She would catch the morning's first train down to the other side of the country, lobby officials to impel the local inspector to visit the plant and measure emissions, and then catch the last train home late in the evening. One outcome of the visits was that the Director of the Inspectorate promised Jennie a high-performance camera to help her with the ongoing monitoring of the coke works. As the leader of the protest group, Jennie also met with delegates from British Coal to find a solution to the long-running row. The talks were constructive and the company's managing director agreed to look into the issues that she outlined in further detail.[196]

Unfortunately, despite Jennie's industriousness, results were not forthcoming. The inspectorate's promised camera never materialised, and no tangible improvements to pollution management at the plant came about. Jennie soon realised that the officials that she was meeting with were merely paying her lip service. On returning from one of her trips to the Inspectorate of Pollution, she told the *Gazette*:

> We are sick of being given tea and biscuits. We want action, not words. Children should be playing on the grass outside my window but they can't because of black smoke from the works. We are asking the local council, who represent the people, to come and fight alongside us to clean up the environment. We have no faith in the Darlington Inspectorate of Pollution because they have done nothing.[197]

Something drastic had to happen if the action group were going to be able to convert all the evidence and support into meaningful improvements for their community. In an unexpected turn of events, Jennie was about to get what she wished for.

In the two years since the public inquiry refused planning permission for a power generation station and a 55 metre chimney at the coke works, British Coal had been exploring legal avenues to circumnavigate the ruling.[198]

Soon after the public inquiry, the company realised that they could legally build the power station regardless of the decision from the Department of the Environment, provided that they utilised existing buildings. The desired 55 metre chimney was not possible without planning permission, but as long as they built a chimney under 15 metres high, it would be within regulations. Construction soon began, and within a few months, they had an operational power station that was ready to be switched on.

Not willing to stop there, British Coal also believed that the Department of the Environment had exceeded its powers in making the decision to dismiss their planning application. British Coal's highly remunerated lawyer persisted with the original argument that the power station would enhance the environment by reducing the flaring of gas. It was also pointed out that the Inspector of Pollution had recommended that planning permission be refused on the mistaken ground that the project conflicted with government policy on desulphurisation. British Coal's lawyer explained that the relatively small power generation project being proposed at Monkton Coke Works exempted the site from desulphurisation equipment, as that policy only applied to large power stations.

Realising that British Coal's team was technically correct, the Department of the Environment volunteered a Consent to Judgement to settle the matter and agreed to bear any associated costs.[199] A Consent to Judgement aims to determine a full and binding resolution to a case. In this instance, the Consent was taken to Her Majesty's High Court of Justice. Based at the Royal Courts of Justice in London, the High Court deals only with civil law cases of the highest importance. Having listened to the company's appeal, the judge quashed the Secretary of State's original decision on a technicality.[200] The secretary now had little choice but to reopen the public inquiry in order to come to a final decision. A second inquiry was convened for the end of January 1990 at South Shields Town Hall.[201]

The legal fight was about to begin all over again.[202] Jennie remained determined to represent her community in the belief that British Coal should be prevented from building and operating a generator. Jennie knew what was happening. She knew that the mighty organisation with considerable resources was trying to wear down the under-financed protesters, but she had faith that the action group would find a way to represent the people of Hebburn.

> We were right then and we are still right now. I think by now people have got the message that we will continue until something is done and people in this area can breathe good clean air.[203]
>
> — Jennie Shearan.

As 1989 drew to a close, 'Greenwatch' capitalised on the season of goodwill by asking for charitable donations to Jennie's cause.[204] With the re-opened public inquiry only a month away, the action group were desperate for finance. Ian Bynoe, the solicitor who had given his services for free at the first inquiry, had now moved to London and was unavailable to represent the campaign team.[205] The action group had not stopped monitoring the pollution of the plant since their first legal showdown and had unearthed even more convincing evidence since the last time they had first stepped on the rostrum.[206] To Jennie and her fellow protesters, it was common sense that the all-pervading smoke and soot must have affected people's health over the years and that people's homes and schools should never have been built in proximity to a coke works. It was even clearer to them that no further development should be permitted on the site.[207]

However, the legal system in which they were about to re-enter required solid proof and an experienced professional representative who could match British Coal's well-paid legal team. With such big financial interests at stake, and an unwillingness to incur a second rejection, British Coal hired the most senior level and thus highest paid lawyer in the system. The firm's representative was to be Lord Colville, a Queen's Counsel. Appointed by the Queen, on the advice of the incumbent Lord Chancellor, Queen's Counsel are the rarest breed and highest calibre of legal eagle. Otherwise known as 'Silks', they are distinguished in the courtroom by their entitlement to wear a silk gown, a winged collar, and a horsehair wig. Beyond the pageantry of their Jacobean accoutrements, they have the courtroom privilege of sitting in the front row and having a portable wooden lectern from which they deliver their speech. Their services are extremely costly, commanding a high day rate to prepare the case then appear at a public inquiry.

The hiring of Lord Colville was an unambiguous signal to the action group that British Coal was determined to win this case at all costs. While a solicitor had been appropriate to properly present the action group's case in the first public inquiry, the activists were now going up against a decision that had flowed from the High Court, and a litigator of the highest standard. A specialist in courtroom advocacy and litigation was needed to compete on an equal footing with British Coal's top-class lawyer. The campaigners needed their own barrister.

They were now in a seemingly unfeasible race against time to source the money to pay for a prohibitively expensive litigator. In the fifteen-round boxing match between the featherweight action group and heavyweight British Coal, a significant lack of funds going into the last round was threatening to indomitably curtail the activists' ability to win the fight on behalf of the people.

Jennie becomes Chairman of Tyne & Wear County Council

A female leader in a man's world

The Hebburn
Residents' Action
Group present their
first petition

BUCKINGHAM PALACE

25th November, 1987.

Dear Mrs Lowry,

I am commanded by The Queen to
acknowledge your recent letter with
which you forwarded a petition from
the Residents Action Group signed by
people living in Hebburn and the
surrounding districts.

In accordance with constitutional
practice and at Her Majesty's direction,
this petition has been forwarded to
the Secretary of State for the Environment.

Yours sincerely,

(KENNETH SCOTT)

Mrs J. Lowry,

Protest around Monkton Coke Works

Britain's first smokeless zone choking on coal dust

"Jenny Shewan

Prisoner in my home

MATTIE Harmeson has a breathing problem and needs daily oxygen with the bottle constantly at his side.

He doesn't blame the coke works for his problem, that probably stems from a life time working in the shipyards, but the constant dust and airborne pollution certainly doesn't help the problem.

"If I get short of breath I need to use the oxygen because it is impossible to open a window," he said.

"I feel like a prisoner in my own home.

"You would be surprised at the number of people around here who use oxygen or inhalers, the smell is getting worse every day," said Mattie.

His wife Nancy tells of her worries for all the young children in the area.

"If you see what the pollution does to the window sills and washing what is it doing to the children's lungs?"

"Mattie Harmeson

From Page 1

should you not breathe if you live near a coke works?" said Jenny.

Thanks to the battling residents it looks as though something will at last be done to clean Monkton up.

The Borough environment has come only two weeks after councillors called on Mr Ridley to intervene in the controversy.

The council's environmental health committee agreed to seek an investigation by the chief alkali inspector after residents filed a complaint under the 1906 Alkali Works Regulation Acts and now the same committee has requested a medical examination of the residents.

The investigation will be headed by Dr Rashir Mallic, one of the region's leading specialists in community medicine.

Campaign

The move has been welcomed by campaigners but the battle is not over yet.

"We have to win this resolved for the benefit of the next generation of not ourselves," said Jenny Shewan.

Next February the campaign moves on to Strasbourg and the European Parliament. "No one will shut us up until this is resolved," said Jenny.

"Clara Green

Mrs Clara Green, is one of the many oxygen users in the area.

"When I saw the photograph of Monkton in the paper I thought yes it looks very beautiful they should come to our house the next morning and see how nice that is after all the smoke and fumes."

"Children play on the grass in the shadow of the works

Health leaflets handed out...

ANTI-POLLUTION campaigners fighting for a clean-up of Monkton Cokeworks are asking thousands of residents to take part in a health survey.

Members of the Hebburn Residents' Action Group are distributing 2,000 questionnaires to people living around the controversial works.

Action group leader Mrs Jennie Shearan said that the questions on the leaflets cover all aspects of health and possible breathing problems.

Campaigners were annoyed when it was decided an official, South Tyneside Council-backed health survey of the area would be too expensive.

Mrs Shearan said: "I hope the people will realise that we can try to get a view of what the position is and get to know whether it's worth going on fighting.

"I think, with the site of Monkton Cokeworks, we will get a lot of questionnaires back."

The questionnaire, entitled Campaign Against Pollution asks whether people think the smoke, noise and dust from the works has reduced, increased or stayed the same since 1981.

Problems

It was then that the number of ovens doubled to 66, although the miners' strike interrupted their operation.

The form also asks about health problems such as arthritis, bronchitis, strokes and cancer, as well as sore throats and sinus and breathing problems. Campaigners hope to discover whether residents believe pollution has caused or aggravated these conditions.

The action group members will afterwards collect all the information together and collate their findings on the possible health problems caused by pollution from the cokeworks.

Getting ready...

PREPARING leaflets for house to house distribution in Hebburn, left to right, Mary Cork, Jennie Johnson, Marian Carey, Barbara Burns, and Jennie Shearan.

WATCHDOG Presenter Lynn Folds-Wood, left, with Jennie Shearan, during filming at the Club Hastings, Hebburn.

BBC examines Monkton fears

Watchdog team talks to residents

THE PLIGHT of Hebburn residents living beside the controversial Monkton Cokeworks will be seen on national television next month.

The BBC Watchdog team were in Hebburn yesterday interviewing angry residents who claim their lives are made a misery by smoke and fumes daily emitted from the cokeworks.

By KATE CORR

The team is also carrying out its own scientific investigation to discover the level of sulphur dioxide in the area.

Watchdog presenter Lynn Folds-Wood said: "I would not want to live next to those coke works.

"I have a family and af-

ter hearing what the scientists have told me I wouldn't want to bring up my little boy living next to them. It's rather worrying."

The team filmed residents in the Club Hastings on the Lukes Lane Estate telling their own horror stories about life next to the cokeworks.

While the cameras roll-

ed a backdrop of billowing black smoke could be seen out of the club's window.

Health problems were discussed, including stories about children on inhalers and suffering from constant sore throats.

Their filthy living environment, where windows must remain firmly locked and where homes are filled with soot and dust was de-

scribed in detail by the furious residents.

Campaign organiser Jennie Shearan, who wrote to Watchdog about a month ago said: "This is the next stage in our fight. The people of Hebburn have been telling their own stories right from the heart."

Vice-chairman Mrs Barbara Burns added: "We are getting no support from our own council — they said they didn't have the money to carry out the research the BBC have been doing — so this publicity can only be beneficial."

Baby coughs up dust

A BABY, only 26-days-old, coughed up real dust after sleeping in his pram for half an hour outside his aunt's home near the Monkton Coke Works.

Baby Michael's mother, Mrs Lorraine Fay said she left her son sleeping and when she went back to him he was covered in black specks of coal dust.

"When I took his hat off I noticed dust in the eyes and nostrils. Later, when he was fed, he coughed and I noticed black specks in the mucus, I was startled," said Mrs Fay.

She then brought her son to visit her sister, Mrs Jeanette Branson, who lives in Wheatsheaf Crescent, Hebburn, close to the Monkton Cokeworks, whose residents have been fighting to clean up for two decades.

It was first recently announced it would be spending £400,000 on dust and grit trapping equipment to clean up the atmosphere.

South Tyneside Council planned to investigate the health of local residents, but after learning that the investigation would cost £100,000 have been forced to refer it back to the Ministry's Inspectorate of Pollution.

A British Coal spokesman said: The cokeworks are an anti-pollution standard reaches the rest of town blowing onto the atmosphere.

Baby Michael's aunt, Mrs Branson said: 'It is a disgrace that babies can't be put out in their prams for some fresh air for a few day without this happening.'

Mrs Jeanie Shearan, leader of the group of campaigners said that the incident highlighted the pollution problem.

She is pushing for an investigation to prove a positive link between emissions from the works and chest complaints in local residents.

★

(Pictured) Jeanette Branson with her 26-days-old nephew Michael Fay, with Monkton Coke Works in the background.

Bringing the community together

Fundraising for the Hebburn Residents' Action Group

Jennie collecting soil samples for laboratory analysis

Jennie and Jenny Johnson outside Monkton Coke Works

Coke works: Now a call for medicals

By John Alevroyiannis

The Hebburn Residents' Action Group leaflet to drive attendance at the first public inquiry

HELL-HOLE!

Cokeworks blasted over Thursday's emission...

MONKTON COKEWORKS ... Thursday's emissions have been slammed by the local residents' action group leader.

SMOKE spewing out of Monkton Cokeworks turned the nearby village into a hell-hole, it was claimed today.

Former councillor Mrs Jennie Shearan said that the alleged pollution from the works on Thursday was one of the worst on record.

She said: "It was just like a hell-hole with the clouds of black smoke. It was disgusting all day long and I couldn't speak because my throat was burned with the smoke."

Story: TERRY KELLY

Mrs Shearan, who has led the fight for a clean-up at the works, and successfully fought a scheme for a power generation station at the plant, contacted both the Alkali Inspector's office at Darlington and environmental health officers at South Tyneside Council.

Speaking from her Mill road Avenue home, right next to the works, Mrs Shearan added: "There was black smoke coming out of all the vents and when things go wrong. It was disgusting and I was literally sick because of the smoke."

She has taken one slide of yesterday's incident.

A spokesman for Monkton Coke Works said they were not aware of any special problem with smoke from the plant.

Mr Ian Rutherford, an environmental health officer on pollution for South Tyneside Council revealed that complaints from residents living near Monkton Coke Works are received on a fairly regular basis.

Pollution at Darlington, which is normal practise.

Mr Rutherford said that his department receives, on average, about three or four complaints a month in relation to Monkton Coke Works.

He added: "At the end of the day, it depends on activities within the works, but every complaint is reported."

Pollution problems can be created when there are ground level emissions from the works, he said.

Action group leader Jenny hits out

HEBBURN residents' action group leader Mrs Jenny Shearan has accused local councillors of failing to support the fight against alleged pollution from Monkton Cokeworks – but councillors say they do care.

The action group is currently distributing a health survey to 2,800 homes in the area. Mrs Shearan said that, unlike concerned residents, councillors have not bothered to return their questionnaires.

She said: "Councillors have refused to fill in the survey that we are sending round. I am amazed. We would like them to help us, instead of us penny-punching, and get this survey done.

"We are doing the work they should be doing. We have put out 1,900 forms and had 800 back – it's just fantastic, this town and the people round here are so concerned what's happening to them."

Hebburn Councillor Mary Fairley said: "I have lived here myself for 35 years – we know the pollution of the coke works. Hebburn advisory committee turned down the application for an increase in emission and for the chimney to be built.

Decided

"I wish people would remember that we want to a public inquiry and it cost the council thousands of pounds for that inquiry which was instigated by Hebburn councillors.

Mrs Fairley said that the health authority – which decided it could not afford to survey Hebburn – did carry out a borough-wide health survey last year.

She said: "That went to every household in South Tyneside and anyone can see that."

→ See Page 19.

Coke works: Call for an inquiry

THE ACTION group calling for a clean-up of Monkton Coke Works is demanding an official Government inquiry into alleged pollution from the plant.

Mrs Jennie Shearan presented the written demand from the Residents' Action Group to Coun. Martin Lightfoot before a full meeting of South Tyneside Council yesterday.

In a statement the group said: "In recent months, the pollution which adjoining residential estates have had to suffer from the Monkton Coke Works has been worse than ever before. Numerous complaints have been made to the South Tyneside Council and to the district Pollution Inspector, but residents have noticed no change in the practises of National Smokeless Fuels, nor the pollution which results."

Complain

Now, the action group, of which Mrs Shearan is the leader, is demanding an inquiry under the terms of the Alkali Works Regulation Act 1906, formally requesting South Tyneside Council to complain to the Secretary of State for the Environment about the pollution problem.

The action group say that the Pollution Inspector has previously served a notice on NSF because the works did not meet current legal pollution control standards.

Campaigner fights for future

A CATALOGUE of broken promises has increased the misery for residents living next to Monkton Cokeworks, a clean-up campaigner has claimed.

Mrs Jennie Shearan, of Melrose Avenue, Hebburn, was speaking yesterday at the reopening of a public inquiry into planned expansion at the plant.

She claimed that during the last two years pollution from the coke works had got steadily worse.

Residents have had to carry out a health survey themselves because South Tyneside Council claimed it could not afford to do so, despite an earlier promise that

Jennie Shearan

it would, she added.

"I arranged a meeting with our local MP Don Dixon who attended a Residents' Action Group meeting on September 8, 1988.

"He listened to what the people said, and stated that he would be in touch again soon, but we have heard nothing, said Mrs Shearan.

"I do not think that anyone should have to put up with the noise and pollution that we suffer. The price that everyone has to pay is too high a price.

"It is not a case of will anything be done – it is for the health of future generations of this area that something has to be done."

£500 donation to coke works campaigners

WOR KATE TO THE RESCUE!

Mrs. Shearan...fighted with the donation from Catherine Cookson and her husband Tom.

CATHERINE COOKSON has donated £500 to help residents living beside the controversial Monkton Cokeworks take their fight for a pollution-free environment to Europe.

Led by former county councillor Jenny Shearan, of Melrose Avenue, Hebburn, three representatives will be flying out to Strasbourg on February 12 to make their complaint to the European Parliament.

But it wasn't until they received a £500 donation from Catherine Cookson that the trio could afford to go.

Mrs. Shearan said she is obviously delighted that Catherine Cookson and her husband have shown their support in this way. "I went to her and sent her a photograph of the view from my home.

She said she couldn't believe it, and wished us all well with our fight to clean up our environment."

Mrs Shearan claimed that the enemy has come and to time. She said: "This weekend the neighbourhood has been ankle deep in dirty. We just couldn't get away from the sand and the dirt coming from the cokeworks. Even if it's the high winds that have been the cause, I even had to wash the sitting room carpet yesterday to try to get rid of the dust and fumes.

Appalling

"We just hope that the European Parliament will listen to us, because no-one else will. All we've got from the local council is but sod sympathy.

"It will be taking evidence of past living conditions with us, including videos, photographs, a document with a street level recording and even some black soot and coke from our homes, to show the appalling conditions we have to put up with."

Residents have been battling for years to reduce emissions from the cokeworks but say matters have improved little.

Ridley rules out power scheme

We've won the fight

Action group in party mood

MRS JENNIE SHEARAN

RESIDENTS on Tyneside were planning a street party to celebrate victory in their long-running battle with a cokeworks.

Their David and Goliath tussle ended when Environment Secretary Nicholas Ridley supported their objections to plans to build a "mini power station" near their homes in Hebburn, South Tyneside.

Supported

National Smokeless Fuels wanted to build generators at Monkton cokeworks, but their planning application was rejected by South Tyneside Council and a planning inquiry was held in November.

The planning inspector supported their case and now Mr Ridley has ruled in their favour.

Jubilant householders plan a massive street party to celebrate their success. Delighted Jennie Shearan, 58, former chairman of the Tyne Wear County Council, said: "It's a case of the little people beating big brother."

Mrs Shearan, of Melrose Avenue, Hebburn, helped launch the Hebburn Residents Action Group which spearheaded the anti-generator campaign along with other residents went to the inquiry, pressing strongly about the plan.

Pledge

"It's a great victory that we have won. It's absolutely great and we'll be having a big celebration party," she said.

But Mrs Shearan pledged to continue the fight. "We won't stop until the air around here is clean. We're fighting for the health and the environment of the youngsters of this town. We want to put and of the soot and muck and dirt which is around, around here.

Cokeworks expansion rejected

RESIDENTS protesting at pollution from a cokeworks were celebrating last night after a Government inspector dismissed a planning appeal to expand the plant.

The Department of the Environment refused permission to the Monkton cokeworks, Hebburn, which it said already caused a high degree of pollution and disturbance.

An application by National Smokeless Fuels Ltd to build a new plant generating sulphur and gas turbine was also dismissed following a public inquiry last December.

Protest leader Mrs Jennie Shearan, of Hebburn Residents' Action Group, praised Environment Minister Mr Nicholas Ridley for blocking expansion of the cokeworks.

SOUTH TYNESIDE
Courier

South Shields, Jarrow, Hebburn & Boldon Edition **AFN** February 9, 1989 Issue No. 23

Coke works protestors take their case to Strasbourg

FIGHT GOES ON...

THE COURIER goes to Strasbourg this weekend with a deputation from pollution scarred Monkton travel by coach and ferry to meet Euro MPs and Commissioners.

The three hope to push for grants to enable South Tyneside to clean up the Monkton Coke Works area and a copy of the Courier is part of their ammunition.

"On the most basic level residents need help to shut out the dirt and the noise from their homes," said Jenny Shearan, former Tyne and Wear county councillor and leader of the protest.

●Pictured left to right, Jenny Shearan and Jenny Johnson outside Monkton Coke Works

"The dust gets into every corner of our lives, our food and our lungs. We can taste it in our mouths constantly," she added.

Turn to page 3

Gazette, Monday, February 20, 1989—11

Cokeworks campaigner reports on trip...

Worthwhile pilgrimage to Europe

By JUDITH DUNN

A TRIP to the European Parliament may have helped Hebburn residents in their fight against pollution.

Members of the Hebburn Residents' Action Group went to Brussels and Strasbourg to lobby Euro MPs on levels of pollution emitted by the Monkton coke works at the area around their homes in Monkton Lane Estate, Hebburn.

JENNIE SHEARAN

And campaign leader, Mrs Jennie Shearan, described the trip as a "worthwhile pilgrimage" shortly after she returned home on Saturday afternoon.

Mrs Shearan felt that members of the European Parliament had listened to presentations by both herself and two other campaigners and were willing to help the residents in their fight against pollution of the local environment by the Hebburn plant.

Mrs Shearan said: "I think everyone in the European Parliament now knows about the Hebburn residents' fight to be able to breathe clean air.

"We put our points across to many members of the European Parliament, including Stephen Hughes, the Durham Euro MP and European spokesman on the environment and Gordon Adams, who is on the Energy Committee at the Parliament.

"Stephen Hughes thinks that we have a case and we will be sent the proper papers and petitions to fill in soon and we will have to send them back to the European Parliament along with our evidence."

Mrs Shearan hopes to not only get the European Parliament's Environment Committee to look into pollution omitted from the works, but also the Energy Committee, who have given grants to National Smokeless Fuels who own the plant.

She said: "We want the Energy committee to look into pollution from the coke works to see that emissions meet European Environmental standards before they give NSF any more grants.

"We are also fighting for closer monitoring of the plant, more capital expenditure on the environment in and around the plant and compensation to residents who are affected by the sulphur emissions and the black rushes, the dust which is produced when coke is pushed out of the ovens before it is cooked.

"We hope to get grants for a health survey and money to help both the council and NSF to clean up the environment from the EEC, as well as double glazing for residents who are bothered by noise from the works.

Mrs Shearan hit out at South Tyneside Council, the alkali inspectorate and NSF who she claims have done nothing to help the residents to a cleaner environment.

She said: "I am annoyed at South Tyneside not giving us a grant to go to the European Parliament as we are fighting for local people and we are trying to get more money for South Tyneside.

"The council has done nothing about the coke works and they always refer us to the alkali inspectorate, who in turn say it is nothing to do with them and refer us back to the council.

"The announcement on Fridays by Coal Products Ltd to spend £400,000 on installing coke side grit treatment equipment seems to be coincidental with our trip to Strasbourg.

"The equipment should have been installed in 1981 when the new ovens were opened and it seems to be just a measure to try to shut local residents up about the plant."

The Hebburn Residents' Action Group visit the
European Parliament

Jennie with her grandchildren

PART THREE

PEACE AT LAST

CHAPTER EIGHT

THE WHITE KNIGHT

'I've done everything I can. YOU do something!'

Jennie had four weeks to find the money to pay for costly legal representation for her beleaguered community in Hebburn, and then bring the lawyer up to speed on the intricacies of the case. South Tyneside Council were once again unwilling to contribute funds to the cash-strapped collective. Their reasoning this time was based on recently introduced government legislation around supporting campaigns.[208] Jennie turned to her Rolodex. Call after hopeless call, the response was consistent. No one had the means to meaningfully bolster the action group's shoestring budget and most people were winding down for the Christmas break. There was too little time to help her. As Big Ben struck midnight and its melodic bongs rang in the New Year, it had become apparent that there was simply no way that Jennie could compete on a financial level with British Coal. The energy giant was effectively putting Jennie's access to justice out of reach.

Then Wendy called.[209] Dr Wendy Le-Las was a 48-year-old environmental campaigner and academic who lived in the cathedral city of Canterbury. Situated in the South East of England, the verdant surroundings in which she was based were in sharp contrast to the industrial complex that Jennie was confronted with each time she looked out the window. While studying for her doctoral thesis on the Greater London Development Plan inquiry 1969–71, Wendy had gained a deep knowledge of environmental procedures and policies. Through her research, she had worked out the techniques required to put forward and defend a public inquiry case around planning reforms, and went on to publish a book on the topic, *Playing The Public Inquiry Game*. Towards the end of 1989, Wendy had been working on a land use planning case in the north of England, and her client was well aware of Jennie's current predicament. On learning more about the matter from her client, Wendy instantly identified with the grossly underfunded action group's plight.

When Wendy was nine years old she had witnessed her parents lose a court case against the Royal Air Force. Her parents should have won but could not afford a lawyer to represent them. Meanwhile, the Royal Air Force had a Queen's Counsel as their legal representative. The subsequent loss had dire consequences for Wendy's family and left them in financial ruin. For two years, the family was effectively homeless, and it took ten years for the family's finances to recover. Within a few years, Wendy's father had died of cancer. These events left a lasting mark on Wendy. She had learned two lessons from this: 'justice' was only for those with deep pockets, and losing a case could have long-term implications.

As Wendy grew older, she embarked on a career devoted to helping ordinary people fight their corner in a world where lawyers were adept at twisting the truth to make white, black, and black, white. Once Wendy found out about the extent to which the Hebburn Residents' Action Group had gone to fight for their community's right to clean air, she was on a mission to avenge the injustice that had been inflicted on her own family. Wendy felt impelled to phone Jennie and see how she could help.

Wendy and Jennie immediately struck up a rapport. They were two warrior queens who had been fighting social injustices their whole life. They deeply cared about the environment, they were strong and they were dynamic. They were kindred spirits. Despite only conducting their initial conversations over a series of phone calls, Wendy thoroughly understood the magnitude of the case, and the mounting urgency to find a representative who could counter Lord Colville. She recognised that Jennie would play a key role in the public inquiry, as a brilliant witness capable of supplying information from all manner of sources. However, she was deeply concerned about Jennie's lack of legal representation and a fast approaching deadline, and strongly believed a very senior and qualified representative was needed for the Hebburn Residents' Action Group.

A barrister was a prerequisite for the action group if Jennie was to avoid being bulldozed by British Coal. Since the publication of her book, Wendy had fought and won four small cases in various parts of England. She found herself on the council of the Environmental Law Association and was becoming acquainted with lawyers who cared about the environment. Throughout the first week of the New Year, Wendy frantically called every single barrister she knew from her home, pleading with them to represent the action group's case. When they heard that the case was intricate, imminent, and they would not even get paid expenses, there were no volunteers.

With each rejection, Wendy's despondency grew. She was fast running out of options. As a new week began, and with just ten days before the public inquiry was set to commence, Wendy feared that her search had become futile. Despite all her efforts, no support had been forthcoming. As in other times of desperation, Wendy got on her bike and rode into the walled City of Canterbury, dominated by its magnificent Cathedral. The leaden skies of that January morning reflected her mood as she rode up and down the steep hills. She could not bear the galling prospect of the Queen's Counsel being paid a huge fee to argue the case for increasing the poison dose to the residents of Hebburn. Wendy walked through the side entrance of the Cathedral, towards the mysterious Crypt under the Quire. Once there, she lit a candle. Hers was not a reverent prayer but a silent scream of rage. 'I've done everything I can. YOU do something!' She did not expect a reply, but it was therapeutic.

On her return home from her five-mile round trip, it occurred to her to ring an experienced solicitor friend in London, who may just know a suitable barrister. Indeed, he did. He immediately recommended a debonair senior barrister named Charles Pugh, who happened to be interested in the practical application of European air pollution law, an area of knowledge that would be central to winning the case. Wendy promptly called him and made the request for his representation.

In 1990, the principles and application of European law were relatively new in the English legal landscape. Many lawyers, judges, and indeed the government had not yet been exposed first hand to its intricacies. The Monkton Coke Works litigation represented an uncharted opportunity for Charles to implement the doctrine of 'direct effect', whereby the Hebburn Residents' Action Group could seek to enforce the rights granted to them under European law. The groundbreaking case would therefore be one of the first opportunities to use European law to challenge decisions made by the State. Notwithstanding the shortage of time and money, Charles replied straight away that he was happy to do it, free of charge. To Wendy's utter amazement, Charles had said yes. She and Jennie could hardly believe it!

At a Christmas drinks party in Dorset a few weeks previously, Charles had met Philip Mead. Charles's mother-in-law lived in the neighbouring village to where Philip's parents lived. Philip had recently completed a Masters in European Law at the European University Institute in Florence, and was beginning his journey to the Bar. Having just started his first six months of pupillage as a barrister-in-training in Brussels, Charles invited Philip to join him at the hearing of the planning inspector to get work experience. So now Jennie had not one, but two legal experts with complementary skills, to fight the case. It was an extraordinary turn of events. By some miracle, four people, scattered across the country and totally unacquainted only a month beforehand, had come together at very short notice to give the Hebburn Residents' Action Group a fighting chance.

The relief that Wendy and Jennie felt was quickly replaced with a sense of panic. There was a pressing need to get the plentiful evidence and paperwork that Charles required, from 1 Melrose Avenue over to Lincoln's Inn in central London. The Honourable Society of Lincoln's Inn is a professional association for barristers and is recognised as being one of the world's most prestigious professional bodies of judges and lawyers. Constructed in the 1500s, the 11-acre complex houses a Great Hall, chapel, and library in which the members have offices out of which they can work. Once Wendy informed Jennie of the incredible news that Charles had committed to the cause, Jennie now faced the logistical challenge of getting the documents over to him with no funds to do so.

Lack of money had not stopped Jennie before, and it would not stop her now. What she lacked in fiscal means, she more than compensated with her ingenuity. Rail Express was a service operated by British Rail that used passenger trains for transporting parcels between passenger railway stations. This service would take a day or more to deliver a consignment. The company also offered a premium service called Red Star, which guaranteed an expedited same-day delivery, and offered last-mile delivery of a package to the final destination, rather than to the nearest train station. Jennie needed to get the documents to Charles fast. Every hour counted. With no connections to anyone within the company, and merely a phone with which to persuade them, she convinced Red Star to deliver the necessary documentation to Lincoln's Inn for free, as a publicity stunt. Within hours, the paperwork was on its way to London, and the *Gazette* was preparing an article focusing on Red Star's support for the cause. It was an audacious move, but it worked.

Furnished with the necessary information, Charles and Philip set about constructing the case for the public inquiry. Lord Colville had been afforded months in which to assimilate the information into a case for the developer, British Coal. On the other hand, the pair's preparation time was minimal. Together, they worked intensively day and night to produce a proof of evidence and the supplementary documents required for submission ahead of proceedings. Under normal circumstances, the summaries were supposed to have been submitted several weeks before the commencement of the public inquiry. Although Philip knew little about being a barrister, he had trained in Brussels and had an even stronger grasp of European Community law than Charles. It was a prodigious amount of work for the pair to mount a case of this complexity in such a short space of time.

As the end of January approached, Charles and Philip visited Hebburn to see Monkton Coke Works for themselves and meet the Hebburn Residents' Action Group. For the professionals, site visits were an important way to transform a two-dimensional paper exercise into a three-dimensional human story. However, Monkton Coke Works was in a league of its own. As Charles and Philip approached the stinking site, they were astonished by how close the operation was to the council houses. Charles had dealt with cases involving hazardous waste incinerators, nuclear power stations, oil refineries, and scrapyards, but he had never seen such a striking example of a polluter placed so close to residents. For Charles and Philip, the plant was the very definition of a 'bad neighbour'. Beyond the extraordinary sight of Monkton Coke Works and Monkton Lane Estate in such proximity, it was eye-opening to talk with the residents who had suffered for decades.

Witnessing at first hand the polluting ovens, smelling the foul air, and engaging directly with the community, furnished Charles and Philip with all the motivation and energy they needed to carry them through the remaining preparation for the inquiry. Jennie and Charles quickly struck up a collaborative working relationship. Although the pair were hamstrung by a lack of time, they had plenty in their favour.

Whilst not a professionally qualified expert witness, Jennie was an unassailable expert in her own right. By now, she had experience in the courtroom and had acquired an unrivalled depth of knowledge on her specialist subject. The all-encompassing data that she had collected, together with the years of research that she had embarked on, combined to form an impermeable story. From her work on the council to her involvement in the first public inquiry, Jennie was also uniquely familiar with the complicated network of bodies that constitute central and local government, and their associated procedures. Moreover, after a lifetime of representing her community, she was completely at ease with marshalling substantive arguments against an expansion to Monkton Coke Works, and able to share the most salient points to Charles on why the plant needed to be cleaned up.[210]

Charles was a highly competent barrister who was well-acquainted with public inquiry procedure and etiquette. Furthermore, his knowledge of EEC environmental protection legislation enabled him to see the case through a broad lens. The aims of the EEC on this matter were to improve air quality throughout the

European Community, by controlling emissions from polluting sources and integrating environmental protection requirements. Over the course of several years, Charles had become acquainted with the intricacies of all the EEC's directives on this topic and could put this knowledge to precise use. He was therefore uniquely placed to detect the key issues upon which a case would hang, and swiftly select the most germane facts and materials that Jennie had put forward to him.

Consequently, in an impressively short space of time, Jennie and Charles were able to organise themselves. Jennie's collection of evidence perfectly complemented Charles's consummate understanding of European environmental law, and he was able to devise an elegant framework that would acquit themselves well in the eye of the inspector. The key issue that underpinned the whole case was the harm to the public interest that further development at Monkton Coke Works would cause should anti-pollution measures not be put in place. This would form the general strategic thrust of his case.

The first day of the public inquiry arrived. In the hour leading up to the morning's first session, several members of the community had gathered outside the entrance of South Shields Town Hall. As they waited for the venue to open its doors, the founding members of the action group and their family members mingled with the journalists who had reported on the issue for years. Jennie and her team were tense but not over-awed. After all, they had been here before.

With the clock ticking towards the start of proceedings, they noticed that Charles and Philip had yet to arrive. As the clock struck ten o'clock, Jennie had little choice but to enter the Chamber, without her legal team.

Once the representatives had settled into their chairs, and members of the public and journalists had found their seats towards the back of the courtroom, Inspector Charnley made some introductory remarks. He explained that the inquiry was to last four days, that he would be listening to all sides as an impartial adjudicator, and he would be taking copious notes throughout.

Jennie looked at her watch and wondered where on Earth Charles and Philip were. Just as Inspector Charnley was about to conclude his opening comments, they entered. The cavalry had finally come over the brow of the hill to save the day! The pair, who had misread the map of the local area, were greeted with restrained acclaim by the Hebburn Residents' Action Group. While Charles and Philip settled into their seats, Inspector Charnley initiated proceedings.

The appellant presented their case first. As Lord Colville approached the centre of the room, the conviction with which he strode gripped everyone's attention. Delivering his opening words with great *élan*, he immediately established an authoritative presence. Aware that the key objective when appearing at a public inquiry is to influence the inspector, he set about refreshing Charnley with the essential context behind British Coal's case.

He explained that the carbonisation of coal generated combustible gas. The gas needed to be burned at a flare stack, from which by-products were discharged into the atmosphere. The installation of a gas turbine could convert the coke oven gas into valuable electricity that would in turn lead to significant cost savings for the firm. Lord Colville argued that annual emissions of sulphur dioxide would neither increase nor decrease. However, sulphur dioxide which would otherwise have been discharged from the flare stack could instead flow from a taller chimney which would result in a reduction in concentrations of sulphur dioxide. In addition, he claimed that by taking advantage of the excess gas at this site, overall emissions of sulphur dioxide would reduce nationwide.

It was a well-constructed and measured summary that did not deviate from the original case. Lord Colville went on to provide an update on the coke works since that surprising outcome from the first public inquiry. He provided details about the turbine and small chimney that had been installed within existing buildings, in preparation for a much-hoped-for green light from the inspector. He disclosed that significant losses of savings had been incurred by the decision to not authorise planning permission for a power generator, and much flaring of gas had taken place. Finally, he explained that gas desulphurisation equipment was not cost effective, and neither was refurbishing the old purifiers.

Next, it was the turn of South Tyneside Council as the Local Planning Agency. The lawyer representing the council pointed to the government's recent highly publicised goal to proactively reduce sulphur dioxide emissions. This was in turn confirmed by Inspector Charnley 'as a matter of acknowledged importance'.[211] The council agreed that the construction of the 55 metre stack would not change emissions of sulphur from the plant, but they differed with Lord Colville's argument in one important aspect: the installation of desulphurisation equipment was obligatory.

During the hearing, the senior inspector from the Air Pollution Inspectorate gave evidence of the inefficient, disorganised, and underfunded way in which the air pollution from Monkton Coke Works was monitored. It provided the room with a real eye-opener of the extent to which the inspectors lacked the resources and the independence to adequately act in the public interest.

Finally, the Hebburn Residents' Action Group were invited to speak. It was now time for Charles and Jennie. In the lead up to the inquiry, the pair had quickly realised that they were the perfect partnership. The erudite Euro-specialist and the passionate layman were both ready to represent the people of Hebburn.

Charles stood up and introduced his case. His well-spoken confidence was the ideal counterweight to the spellbinding conviction of Lord Colville. Charles's core intent was to dissuade Inspector Charnley from authorising any further development of the plant until anti-pollution measures were put in place. He reminded Charnley that the inspector from the original public inquiry had accepted that Monkton Coke Works had not operated to required pollution standards and were in breach of several obvious abatement measures. Charles contended that the plant continued in its failure to follow best practicable means, and that any assurances that National Smokeless Fuels would better control pollution in the future had no credibility because of the firm's track record.

Charles delivered his mellifluous assertions with flair. Monkton Coke Works had taken no preventative actions to safeguard the community against pollution. The plant's reactions to previous complaints had been consistently muted. National Smokeless Fuels had never once acknowledged the risk to public health that their site posed. The tonnage of sulphur dioxide that the site produced was five per cent of the whole country's total industrial emissions. Local residents had not been able to get assistance from the Inspectorate of Pollution, so had been compelled to take matters into their own hands. It was a shocking list of events that was delivered with alacrity.

Charles was well informed of the absurdity of British pollution control legislation that dealt only with what was incontrovertibly hazardous to human health and not with what was acceptable or unacceptable to people living on the doorstep of a coke works. He knew that unless the facility constituted a proven danger to human health or the environment, that it would be impossible to get legislation put in place that actually closed down the plant. As such, he argued that a medical inquiry into the health impact of pollution in the area was necessary. The two-year study that had been agreed to, then promptly cancelled, would have to be reconsidered. This was one matter that all parties agreed on. Lord Colville, South Tyneside Council, and Charles were united in the agreement that a formal study was necessary, for a formal link to be established between air pollution and damage to health.

With the inquiry having adjourned to the next day, Charles and Philip had overnight to finalise the concluding speech on all aspects of the case, including the issue's relationship to European Commission law. The pair knew EEC legislation on air pollution just as well as Jennie knew the pain that Monkton Coke Works had inflicted on her community.

Over dinner, Charles and Philip discussed the skeleton of the premise that Charles had prepared. Philip, who was by now thoroughly subsumed in the case, persuaded Charles that he could make some significant improvements to the speech. With his fluency in the most relevant articles in European anti-pollution directives, he set about rewriting the argument for Charles. From ten o'clock that evening until the early hours of the morning, the pupil composed a speech that would support Charles's contention that National Smokeless Fuels had infringed the law on emissions.

The public inquiry resumed the following morning. Charles cited heavily from the EEC/84/360 and EEC/84/609 rulings that were summarised in the improved speech, maintaining that the firm had a responsibility to preserve and protect the quality of the environment with pollution abatement measures. He declared that the planning permission that the firm was proposing would 'substantially alter' the plant and, as such, several measures would have to be taken to prevent significant air pollution. The desulphurisation of Monkton Coke Works was an absolute, indisputable priority.

Charles concluded his presentation with the reiteration that National Smokeless Fuel's appeal should be dismissed because their proposal contained no measures for the prevention or reduction of air pollution and that existing equipment at the works designed to curb pollution had fallen into disuse. He emphasised that conditions would have to be imposed should the 55 metre chimney be allowed, including the desulphurisation of the entire gas production at the works before any new scheme was implemented. Finally, he requested regular checks by independent experts on dust and fumes emissions, as well as noise levels.

Charles's forensic cross-examination of European directives had been masterful. The White Knight had spoken. Now it was time for the Great Lady. Working in harmony with Charles's enthralling expedition through EEC directives, Jennie focused her line of thought on the health impact of living next to Monkton Coke Works. She began by saying that for decades she and her neighbours had been concerned about the sulphur emissions and other pollutants that had so liberally emanated from the site. She questioned how the fumes could not be harmful to residents who lived on its doorstep, and remarked that the only development that should take place should be to tidy the plant up. Her argument was boosted by readings from a diary that Jenny Johnson had kept for the six months leading up to the inquiry. Each day Jenny Johnson had diligently monitored and described the pollution levels. The extracts that Jennie read out loud to the courtroom represented a startling contrast to the highbrow approach of her predecessors that day. Barbara then stood up and joined her mother in giving evidence. Barbara's speech was as impassioned and considerate as the presentation she gave during the original public inquiry.

The inspector invited the appellant to address any questions to the Hebburn Residents' Action Group. Lord Colville stood up. He was well aware that the action group had taken his client by surprise in the first public inquiry, and was determined to undermine the activist's case. The grandiosity of his mannerisms clashed with the understated formality of Jennie's attire. Ostentatious ceremony against unpretentious pragmatism. In his unmistakable magisterial style, Lord Colville began to imperiously ridicule everything that Jennie had said. He dismissed the subsidiary evidence of visual aids and diary readings as overly subjective. He derided her claims of the worsening health of residents due to the plant as 'scaremongering'.[212] He also disparaged the unscientific methodology of her health survey.

All is fair in love, war, and public inquiries. Jennie was not done yet. After calmly listening to Lord Colville's scrutiny, she was asked if she had any final evidence to show. Jennie answered that she did, and would need a television and video player. A trolley hosting an enormous television was wheeled into the centre of the room, with the colossal screen in full view of the audience. A VHS cassette was inserted, and Jennie asked the room to watch. Over the following five minutes, a professionally edited video displayed in full colour and sound the degradation that had been forced upon the residents of Monkton and Luke's Lane Estates for decades. The choking fumes. The repugnant smoke clouds. The howling sirens. The petrifying flames. The poisonous puddles. The frightening proximity to residents. Nothing was missed. This was Jennie's *coup de grâce*. Lord Colville was silenced. His face dropped. He looked to his left, and then to his right, to see a room full of people who were repulsed by the inhumane dirt, noise, and pollution of the very plant that he had been defending.

Charles, Philip, and Jennie had given it their best shot. The legal team's unrivalled knowledge of European law had been buttressed with Jennie's compelling visual evidence. As the inquiry drew to a close, Inspector Charnley was taken on a site visit of the coke works. The strong unpleasant smell of the mixed gaseous sulphur compounds greeted him with open arms, and the action group was relieved that he had been able to experience first hand a little slice of what they endured daily.

On returning to South Shields Town Hall for the final afternoon's closing statements, Jennie knew that the next few months would be a waiting game.

By April 1990, Inspector Charnley had completed his report on the re-opened inquiry and sent his recommendation to the Secretary of State for the Environment.[213] He had spent three months digesting the arguments, and the wide-ranging documentation that had been submitted. Academic paperwork such as EEC directives and histograms of annual sulphur dioxide emissions were all analysed. Less empirical, but equally valid evidence, such as a sample of comments from the action group's community survey, Jennie's regular photos of the site, and Jenny Johnson's six-month day-by-day diary of emissions from the plant were also considered.

Following the protocol of such reports, Inspector Charnley began by outlining the individual cases of the appellant, the Local Planning Authority, and the Hebburn Residents' Action Group in a succinct yet thorough manner. From the introductory courtesies to the summaries of the key arguments, and all the way through to the conclusion and recommendation, each distinct paragraph represented a synopsis of a complex particularity of the case. With just over 100 paragraphs, the report was an in-depth recap on which the Secretary of State could base his final decision.

The inspector's report ended with a series of conclusions. The key concern that drove his opinions was the recognition that sulphur dioxide was undoubtedly damaging to the environment and that the protection of the environment was his key duty.

His first deduction was that the power generation station would make no significant change to pollution levels from the plant. He accepted that the generation of electricity at the plant would lead to less coal being burned elsewhere, but noted that the difference in the net amount of sulphur dioxide emitted into the atmosphere would be negligible. He had also determined that while the proposed 55 metre chimney would not worsen pollution, it would not reduce it either.

His second verdict was that the residents deserved a better quality of life. Inspector Charnley acknowledged that while Monkton Coke Works was operating to the normal standards applicable to the coking industry, the fact that it was situated near two large housing estates made the plant a 'bad neighbour'. He wrote extensively about the residents who had suffered a high degree of pollution and disturbance from the plant.

On the topic of whether emissions from the plant worsened the health of local residents, Inspector Charnley supported the overall sentiment that an official comprehensive study was required, while also admitting that it was common sense that such pollution could not be anything other than harmful.

Crucially, the inspector identified that the installation of costly apparatus to desulphurise the gas would lead to a significant reduction in the emission of sulphur dioxide. He saw this as having a considerable environmental benefit to the local community and further afield.

Having outlined his logic, his recommendation was therefore to allow British Coal's appeal and grant planning permission for the installation of a 55 metre high chimney, on the condition that flue gas desulphurisation equipment was fitted, regardless of the expense that would be involved.

The news was simultaneously disheartening and encouraging for Jennie. The thought of Monkton Coke Works continuing to expand made her deeply nauseous. On the flip side, Jennie reminded herself that the requirement to install desulphurisation equipment could no longer be dodged by British Coal. As with most compromises, neither party was entirely satisfied. British Coal had been granted the profit-driving planning permission that they had coveted for years, but with the caveat that they had to invest in the expensive equipment that they had been avoiding for nearly a decade. The Chairman of British Coal had been tasked to drive cost reductions and increase productivity. This ruling enabled him to drive productivity but he would need to add significant costs to his balance sheet. Meanwhile, although the action group were pleased that anti-pollution equipment was finally going to be introduced to the plant, they were deeply upset by the prospect of the coke works continuing to expand.

As spring turned to summer, some encouraging news emerged.[214] The public health probe that had been promised to residents two years earlier and then just as quickly swept away would be carried out after all.[215] Monies had become available from South Tyneside Council from underspending in parts of their budget, and they were now able to cover the £50,000 cost. While the action group strongly believed in their own health survey results, they saw immense value in an official study. A comprehensive investigation conducted by qualified scientists would either legitimise their own findings or put the community's worries about the long-term health impact of the coke works to rest. Jennie publicly pledged to support the scheme and started to work closely with the medical professionals overseeing the study.

> We have provided a great deal of information for the scientific team and feel sure they will do a thorough job.

We will abide by their findings. If they find the works
to be totally safe for residents living in its shadow then
that is all well and good. But if they find the opposite
it will show that we were right all along.[216]

— Jennie Shearan.

Countering the action group's cooperative approach, the Inspectorate of Pollution once again refused to back the initiative,
maintaining that despite the closeness of housing to the plant
it did not feel any health hazard existed. The inspectorate referenced several recent investigations that had been made around
the world that revealed no statistical evidence of any adverse
effects on the workers who were exposed to emissions.

With or without the inspectorate's backing, the research
was going ahead. Dr Rajinder Bhopal from the Department of
Epidemiology and Public Health at Newcastle University was
assigned to lead the project. Their chief focus was studying the
pattern and causes of health-related events in specified populations, and they set out to establish a link between pollution
from the plant and the health of residents nearby.[217] The work
that they were doing was the first of its kind in the region and
had the potential to become a valuable cornerstone for all future
complaints of a similar nature. As such, they resolved to carry
out the study impartially and thoroughly.

With the increasing concern about the environment
and its health consequences, similar problems will arise
in the future and will need investigation. At present,

the methodology for investigating complaints of this nature is poorly developed. The present problem allows us to develop such a methodology. Secondly, it seems unlikely that public anxiety will be assuaged by the reassurance that the chemicals emanating from the coke works have no known toxicological hazards. Nothing less than an empirical demonstration that this is the case will suffice.[218]

— Dr Rajinder Bhopal.

As the doctors' ultimate goal was to answer the question of whether the population living closest to Monkton Coke Works suffered more ill health than a similar population elsewhere in the borough, they had to approach the challenge in a rigorous manner. They landed on a five-step process in order to reach a robust verdict. They analysed deaths, registrations of cancer and hospital admissions among people living close to Monkton Coke Works compared with people who lived further away. They also compared permanent sickness and disablement rates, as well as low birthweight babies. Penultimately, they looked into what kinds of illnesses were being reported to doctors, and how often they occurred. Finally, they developed a survey for residents near Monkton Coke Works and those in a control group. The pioneering project was going to take two years to complete, with over 4,000 questionnaires set to be distributed across the housing estates closest to Monkton Coke Works in November 1990.[219]

In the weeks prior to the questionnaires being distributed across South Tyneside, the Chairman of British Coal made a sudden announcement. The firm was to decommission Monkton Coke Works.[220] British Coal pointed to the downturn in the coke market and the severe contraction of demand from the decline of the steel industry as the factors behind the decision.[221] It was jaw-dropping news for the action group. While the officially stated reasons for the shutdown of the facility were economic and not environmental, the campaigners had their own theories about the snap decision.

The action group knew that the firm was under no obligation to give a fully truthful reason behind the closure. Given that British Coal had just invested so much time and money into expensive representation at the public inquiry to expand the plant, the activists were convinced that the firm had been motivated by forces other than macroeconomics. The ticking timebomb of a coke works operating next to a housing estate had finally exploded.

The action group inferred that British Coal had been unwilling to invest in the stipulated desulphurisation equipment. They imagined that the expensive pollution abatement measures that the firm had been so reluctant to install, which was now an imperative, would have proven too costly.

The activists also believed that the health report would reveal some damaging insights that would reflect badly on British Coal. Their conjecture that the prospect of a damning investigation from the epidemiologists was a material factor was also backed up by the local press. With the vacillating saga inspiring ever more witty headlines, one article titled, 'It's a case of no smoke without ire', openly suspected that the imminent health study results must have been part of the equation.[222]

When Charles Pugh and Philip Mead heard the news, they described the decision to shut down Monkton Coke Works as an unconfirmed direct hit. It was as if the public inquiry's verdict had been a moment of anagnorisis for British Coal. Weighing up the ever-decreasing earnings that they would accrue from a steadily lowering demand for coke, against the significant expenditure required to pay for the pollution filters, as well as a potentially incriminating health survey, British Coal's decision must have become inescapable.

For Jennie it was a moment of retribution. It had become clear that there was no pollution abatement system in existence that could have made these homes decent to live in. The moment had arrived that the majority of Monkton Lane Estate had often feared would never happen. It was a seismic shift from months earlier. The idea of the works closing had seemed an impossibility, and now it was a reality. So much had happened in such a short space of time. With the Secretary of State's insistence on desulphurisation equipment, then the authorisation of an official health study, and now the imminent closure of the plant, it felt as if the house of cards had finally tumbled. When *Look North* televised an interview with Jennie, questioning her on the news, she was visibly exultant.

> The thought of being able to come out of my front door and to breathe clean air … you know, we thought it would never happen. I think it's great.

It's very sad that the men are going to lose their jobs, but I think that there is nothing but good that can come out of it. The site can be flattened, all the polluting soil can be taken away, and better jobs, more cleaner jobs can come in their place.

Most of all, the people that live here can be considered.[223]

— Jennie Shearan.

Where lesser mortals may have taken their foot off the gas at this stage, Jennie switched into fifth gear. Although she had never played chess, her ability to think several moves ahead could have made her a Grandmaster. Her attention turned to how the site could be repurposed after half a century of hazardous waste had infiltrated its soil. If the area was to be safely redeployed it would be crucial to ascertain its suitability for future development.

Jennie organised a visit to meet David Trippier, the government's environment minister, to persuade him to authorise a subterranean survey of pollution effects at the plant. When booking the train tickets to take herself, Barbara, and Jenny Johnson down to London, it dawned on her that the trip could be a good opportunity to invite a fourth attendee who was based relatively close to the capital, but who she had yet to meet in person.

Ever since Jennie had received that fateful introductory phone call from Wendy, the pair had maintained a regular dialogue. From nightly commentary on how the public inquiry was progressing to the joyful proclamation that Monkton Coke Works was to be closed down, Jennie had kept Wendy updated every step of the way.

Earlier in her career, Wendy had done research in a university epidemiological unit, and was familiar with the language of morbidity, mortality, and the difficulties of establishing cause and effect. Jennie thought she would be a helpful addition to the appointment with Trippier. When the pair met outside the inspectorate, just before the meeting began, as it was as if two old friends had reunited. Their mutual respect and shared passion for a better environment had bonded two perfect strangers. As the meeting with Trippier got underway in a small conference room, Jennie produced photos and videos of the facility in an attempt to bring to life just how hazardous the soil was likely to be. She then requested guidance on how to get community compensation from the parties that she saw at fault for the whole saga.

At this moment, Wendy pointed out to the minister that his government would not countenance this injustice happening in a southern town such as Guildford because the inhabitants voted Conservative, but the residents of Hebburn living hundreds of miles from London, voted Labour and were therefore expendable. Trippier mistook Wendy for a medical doctor. All he could say was 'Yes, doctor'. Trippier said that a national survey was being undertaken of the health effects of coke works on local populations. 'The results depend on which questions you are asking', Wendy retorted. 'Yes, doctor', he again replied.[224] The next morning Wendy's astonished husband received a phone call from the minister's secretary giving Wendy permission to check the wording of the questions. Such a survey should have been conducted decades ago.

The quest for justice was to become a topic that Jennie would be go on to publicise in interviews with the local press.

> We do not believe British Coal did enough to prevent the coke works from polluting the surrounding area, and the council went against all advice when it built houses so close to the plant.[225]
>
> — Jennie Shearan.

Her persistent eye to the future extended beyond her own neighbourhood's needs. She felt that the action group's efforts had laid the groundwork for future communities beyond South Tyneside to protect their environment.

> People are more educated now and won't allow something like Monkton Coke Works to pollute the atmosphere again.[226]
>
> — Jennie Shearan.

In December 1990 the 53-year-old plant finally ceased production. Unfortunately, what ought to have been a period of celebration for the action group quickly transformed into dismay. British Coal announced that it would be mothballing the deactivated plant, and not dismantling it. This meant that the firm would be preserving the site and its equipment, in the eventuality that there was an upturn in demand for coke. The plant could start up at any time. On the front page of the *Gazette*, the day after British Coal's statement, Jennie condemned the move.

> I would never trust British Coal. They are just dangling the workers and residents by a string. I think that they will just leave the plant to stagnate and be a blight on South Tyneside. They have never tried to be a good neighbour. The fight is far from over yet.[227]
>
> — Jennie Shearan.

As the go-to for any matter relating to the Monkton Coke Works issue, Jennie was juggling many balls and was as busy as ever. Her days were filled with conversations with government officials, discussions with her confidants on how to secure justice for her community, and interviews with journalists wanting a soundbite for their next column.

The region's main newspaper saw this moment as an opportunity to reflect on Jennie's efforts and dedicated a feature to her endeavours. The opening paragraphs of the two-page spread served as a fitting summary of her character.

Woman of the Week

If you read nothing else in tonight's Chronicle – read this.

Jennie Shearan is a true Geordie and Labour Party activist. Her father was a trade unionist, her role model was Ellen Wilkinson. In the world of local politics, she is a natural street fighter. But she is also a kind and loving woman.

Jennie Shearan has a highly developed sense of what's right and wrong. She also has what amounts to a compulsion to speak out publicly if she believes something is unjust or damaging to fellow humans.[228]

With 1990 drawing to a close, a group that needed Jennie's input was the Department of Epidemiology. The response rate for the health questionnaires had slowed down as residents' minds turned to Christmas preparations.[229] The researchers needed a seventy-five per cent response rate in order to draw well-grounded conclusions, and they had only reached just over fifty per cent.[230] Jennie organised with the *South Tyneside Times* to make a front-page plea to the community to submit the paperwork to the doctors.

> People must realise how important this survey could be, not just to people living near Monkton Coke Works, but also to others living round other coke works in Britain and in other countries. If anyone is having difficulties filling the forms in, either myself or other members of the action group would be more than willing to come along and help them fill out the questionnaires.[231]

— Jennie Shearan.

In a typical act of service to her community, Jennie ensured that the journalist print her home phone number at the end of the article for anybody to contact her directly, should they need support in completing the questionnaire. Response rates shot up, and the doctors got the information they needed just before the end of the year.

Jennie had never done things in half-measures, and she was not about to now. She was going to get the job done. It had been a transformative twelve months for the action group, but there were many loose ends to tie up.

CHAPTER NINE

RISKING EVERYTHING

'He does not have to live under an umbrella of pollution and breathe in filth all day. He should come and see what we have had to put up with. We have had a few months of heaven and we don't want to go back to hell.'

Throughout the opening months of 1991, there was a heightened concern around Monkton Lane Estate that the mothballed plant would resume operations. With every passing week that the plant lay dormant and no updates on its future state, rumours began to fly about a return to the bad old days. The inertia from British Coal aroused suspicion, but it was the government that broke the silence.

In the intervening years since Michael Heseltine had risen to prominence as the Secretary of State for the Environment in charge of implementing Thatcher's 'Right to Buy' legislation, the charismatic minister had become an established, albeit controversial, figure in politics. After a decade of holding various Cabinet positions, championing the regeneration of Liverpool, and unsuccessfully challenging for the leadership of the Party, he had recently been re-appointed as Secretary of State for the Environment. John Major had taken over from Thatcher as the Conservative leader and prime minister. Major saw in Heseltine a political heavyweight who he could rely on to implement his key policies and oversaw Heseltine's reinstatement to the front bench. A critical part of Heseltine's remit was to manage the reform of a contentious system of taxation that Thatcher had introduced a few years prior. The Poll Tax, as it was most commonly known, imposed a single flat-rate tax on every adult, regardless of their income. This initiative was widely criticised as being unfair and burdensome on lower-income individuals and became a symbol to many Labour Party supporters for how out of touch the Conservative Party was with their reality. As the public face of the Poll Tax initiative, Heseltine became the subject of derision by its opponents. When the abolition of the Poll Tax was ultimately announced in March 1991, that disdain compounded.

The Poll Tax debacle made Heseltine front-page news throughout the country. That same month, he took another decision that was to keep him on the front pages of the North East newspapers for a little while longer. In a letter addressed to South Tyneside Council and the Hebburn Residents' Action Group he wrote that he disagreed with the inspectors' recommendation from the re-opened public inquiry. Working alongside Trippier and other members of his department, he had determined that the gas desulphurisation equipment was not essential and that the 55 metre chimney could be built without it. He went on to explain that the plant was operating within air pollution regulations and that the projected emissions would be relatively small in comparison to the cost of the equipment. As such, despite the government's proclaimed policy to progressively reduce emissions of sulphur dioxide, Heseltine decided to overrule the requirement to install desulphurisation apparatus.[232] In one swift signature, Heseltine had given the go-ahead for Monkton Coke Works to become a power station liberated from any requirement to control the sulphur emissions that would be generated. Heseltine ended the letter with an offer to both the council and the action group to challenge his decision in the High Court if they so wished, but that the offer must be accepted within six weeks.

The town of Hebburn was shocked.[233] Nobody had seen this coming. Among the many front pages expressing their disbelief, the *Gazette* was the most direct, with the outcry capitalised, 'YES TO POWER STATION'.[234] The council, who for so long had stayed on the sidelines, were vocal in their disdain. Councillor Charles McHugh, Chairman of the Town Development Committee, summed up his enraged colleagues' vexation: 'It makes a nonsense of the government's claims that it cares about the environment, and it could lead to an even greater health hazard in the area'.[235] Even British Coal had been taken by surprise and stated that despite Heseltine's ruling, the plant would remain mothballed for the foreseeable future.

Nobody was more shocked than Jennie. Her response was ferocious.

We will fight this decision again in the High Court if necessary. It was the residents who took the case to the European Parliament, not the council, and we will do it again. As the Minister of Environment, he's got to practise what he preaches. He got the Poll Tax wrong, and now he's got this wrong. I don't believe Mr Heseltine has been in the job long enough to have even read the papers. He is supposed to be for the environment but has willy-nilly made this decision without a thought. [236]

To me, Heseltine is totally ignoring all the rules on the environment from Europe and we are going to fight this decision all the way. By the laws of the UK and the EEC, British Coal are legally bound to install desulphurisation equipment. Mr Heseltine is flaunting both laws by not making [this] a condition. I know people will not stand for it.[237]

He does not have to live under an umbrella of pollution and breathe in filth all day. He should come and see what we have had to put up with. We have had a few months of heaven and we don't want to go back to hell.[238]

— Jennie Shearan.

Through her visits to Strasbourg and Brussels, and her interactions with Charles Pugh, Jennie had gained an implicit understanding of how she could challenge Heseltine's decision. This was pre-Brexit Britain, and at this critical juncture she would take the case to the European Court of Justice if necessary, to have European law imposed by direct effect in the UK.

As she hung up the phone call from another journalist who had reached out for a quote, the ugly topic of money reared its head. How was Jennie going to take on Heseltine? She had less than six weeks to secure legal representation, and a bank balance that was smaller than miniscule when compared to that of the government. Generous donors and pro bono lawyers had gotten her this far, but she was by now too tired to ride the rollercoaster of fundraising and hunting for an unpaid lawyer. Not only did she not have enough money for a lawyer, she did not have enough money to cover the fees that would be incurred if she lost the case.

She looked around her living room. Framed photos of her family and her time as Chairman of Tyne and Wear hung along-side the war medals that her dear David had heroically earned. Jennie may not have had meaningful amounts of money, but she did have her home. The politician who she was prepared to fight in court was the very man whose 'Right to Buy' scheme had led to her owning the property that she was standing in.

She looked at a photo of her husband and was reminded of his constant words of encouragement and endless delight at her achievements.

Jennie made up her mind. She went to the bank and put her house as collateral to cover the legal costs. She found a solicitor and she crafted her case. In the space of three weeks, she served the papers to the Secretary of State and braced herself for yet another day in court. Days passed, with no reply. Weeks turned to months, and no government response appeared in the mail. Summer turned to autumn, then autumn turned to winter.

In January 1992 the uncertainty was finally settled. The Chairman of British Coal announced that the mothballed plant would be demolished.[239] Monkton Coke Works was to be no more. It was a landmark moment, not just for Jennie, but for the whole town. Profit, not people, was the driving concern behind British Coal's decision, but it was people power that had forced their hand. There was to be no more filth or pollution. It was unbelievable news for the hundreds of residents who had contributed to the fight to clean up Monkton Coke Works. The core members of the Hebburn Residents' Action Group were as jubilant as their neighbours, none more so than Jennie, who remarked, 'It's a great day for the children of this area who can now look forward to clean air'. The anti-pollution campaigners who had remained united for so many years would at last be able to enjoy the fresh air for which they had fought for so long.

There was more news to follow a few weeks later. The health survey results were in, and the *Gazette* was the first to write about it.[240] Their compendious headline said it all: 'Chokeworks'.[241] After two years and £50,000, the doctors at Newcastle University had concluded that residents living on estates around Monkton Coke Works were more likely to suffer respiratory problems than in other parts of South Tyneside.[242] While their findings could not point to any noticeable increase in mortality or cancer rates, the evidence showed that much of the sinus problems and chronic coughs had arisen as a result of exposure to emissions from the plant. The results of the health survey proved what the action group had been saying all along; the pollution from Monkton Coke Works had damaged the health of the community. For Jennie, it was a victory of sorts, in that it proved, without doubt, the harm that the hazardous ovens had inflicted. Her overriding emotion, however, was one of frustration that such an obvious conclusion had taken so long to determine, and cost so much, both in terms of finances and quality of life.

> I didn't need the report to tell me whether or not we had a problem here from the coke works. It is heaven now [that] the works are closed. The action group fought long and hard to have this study done, and perhaps it is too late for us. The damage has been done. Perhaps the study will now help others who have a problem similar to ours, though there is still a lot to be done for the people living around here. The poison is still there in the people.[243]

> — Jennie Shearan.

In the months that followed, Jennie became an authority figure in the North East on a variety of local environmental issues. She would often be invited by journalists to share her thoughts, with her comments consistently preceded with descriptors such as 'Hebburn environmental leader' or 'anti-pollution campaigner'.[244] She was, as always, outspoken and earnest with her opinions.[245] Her reputation even transcended politics; the Liberal Democrats openly praised the efforts of the Hebburn Residents' Action Group for their fight to stop pollution.[246]

The plant began to be dismantled.[247] Lifting back the now pristine lace curtains of her bedroom window, Jennie began to observe the site as it gradually diminished. The fight was over. Jennie could now reflect on the tumultuous journey that she and her activist friends had been on. She had recently entered into her eighth decade, the last two of which had been principally devoted to fighting the worst environmental eyesore in South Tyneside. It felt anomalous to observe the disbanding of a 50-year-old site that had incessantly cleaned coal for smokeless consumption, then dispersed its dirty deposits all over her community.

From her window, Jennie saw the culmination of years of protests by ordinary people who had stuck together and taken on the might of British Coal. People power had moved a mountain. The action group's demand for clean air for the residents in their area had taken them on a long and winding road. They had appealed for environmental justice to as many people as they could think of. They had corresponded with local councillors, Members of Parliament, government health watchdogs, officials at the EEC, and even the Queen. Their fight to curb pollution had been embraced by the local press and then magnified by the national media. At each stage of their battle, they deployed every tactic in the activist playbook to build their case about the harmful effects of the coke works as they pitted themselves against a massive corporation. From petitions to protest marches, rebates to sit-ins, and surveys to videos, the action group had left no stone unturned.

They had fought in unison, and it was Jennie who led the way. She could not have done what she did without her daughter Barbara, Jenny Johnson, Marian Carey, Maudy Cork, and Jenny Lowry, to name but a few. When others grew weary of the fight and feared their campaigning was fruitless, the action group kept the momentum going, and kept the pressure on the authorities. But it was Jennie who had really put it all on the line and given it her all. She had been the relentless mainstay and face of the campaign, who had masterminded every move from her living room and knocked on every possible door.

As she opened her window slightly and breathed in the fresh air, she realised that she had done the unthinkable. She had gathered a group of fellow middle-aged neighbours and tackled an issue head-on that the local politicians would not touch. By orchestrating an uninterrupted flow of publicity against the coke works she had put unforeseen pressure on British Coal, and hastened the demise of the plant. Within the four walls of her home, she had built a movement from the ground up and coalesced a community. In retrospect, the tide of public opinion had become so strong against the works that the closure seemed inevitable, but she knew what it had taken to get there.

Throughout every setback and breakthrough, Jennie had remained dogged in her pursuit of environmental justice in her community. She had believed so strongly in a better future for her town that she had vowed to continue Attlee and Bevan's work to eliminate 'Squalor' for future generations, whatever problematic circumstances she encountered. As she told a journalist in the lead up to the re-opened inquiry, 'It would be the easiest thing in the world to wash my hands of it, but then I couldn't live with myself'.[248] In the course of her all-consuming anti-pollution campaign, she had lost her husband, been derided by some disapproving sections of the community, been rejected and ignored by countless officials, and was consistently let down by craven local councillors who could have more proactively supported her cause. Not once did she give up.

Jennie did it all with heartfelt tenacity. She was all words and all action. Yes, perhaps at times she had been brusque in the manner in which she challenged the lackadaisical local authorities[249] and British Coal, but her down-to-earth attitude was a trait that had won her many allies.

She may not have been able to match the riches of the corporations that she had gone up against, but she had devoted the most precious resource of all; her time. Not hours, not days, not even years. Decades of her life, spent writing letters, being passed from person to person on the phone, collecting petition signatures, amassing evidence, organising meetings, travelling the length of the country and beyond, taking photographs, and recording footage. The matter had consumed her every thought and dominated her life. And now it felt like a chapter had ended.

As she closed the window and drew back the curtains, she looked around the freshly aired room. Silence filled her home.

CHAPTER TEN

SLAYING THE DRAGON

'I know your face, and your voice. You're the woman who gave us clean air.'

The final demolition day arrived for the plant where fires had once burned day and night, with fumes and noise invading the housing estates on the other side of the fence. It was November 1993 and for Jennie, this was a bittersweet moment. She had spent more than half of her life seeing Monkton Coke Works every day as she stepped outside her home. Bulldozers were dotted around the industrial wasteland, with only two prominent chimneys remaining intact. Spread among the 9,000 tonnes of deteriorating coke and coal piles were regional television crews, there to cover the flattening of the works. A 10-year-old local Boy Scout had won a raffle competition to press the button that would detonate the explosives at the stroke of midday.[250] As had long since become customary with any coverage relating to the plant, Jennie was the first person to be interviewed in the news features later that evening. British Coal had not invited her onto the site to observe the demolition up close, but she had joined a group of local residents from Melrose Avenue to witness the end of an era.

There were other interviews with neighbours, who shared mixed emotions, ranging from concern about the loss of jobs, and excitement to see the condemned chimneys come crashing down, but the predominant opinions were ones of relief and hope. As the face of the long-running clean-up campaign, Jennie's reflections immediately after the demolition echoed the majority of her neighbours' sentiments on that day.

> I'm glad the chimneys have come down and no one's hurt themselves and everything's gone off all right. It's sad in one respect. But happy in another. We won't miss them. The pollution was bad, the noise was bad, the smell was bad.
>
> When we first moved in, you didn't know how bad the operation was going to be. Consequently, as time went on, you learned to live with it. It all came out when I went around my neighbours' houses with a petition. I was shocked when I saw how many people were on ventilators and on different forms of oxygen.
>
> I've spent a lot of time outside those gates over the years, and inside them, because I had to go many a time inside at night to complain about the filth and the pollution and the noise. You couldn't sleep at night, and you got no peace through the day.

It was so strange to see them come down after all
the years of fighting. There were hundreds of people
lining the streets to see them come down. Many had
suffered health problems because of the site and it
was a strange feeling. I couldn't help but shed a tear.

Now we can just think of the future of what it's
going to be like for the people in this area. It can only
improve one hundred per cent. It will now be a green
field where children can play safely. This is what we
fought for. A better environment for the future generations of this town.[251]

— Jennie Shearan

After over 50 years of living in the shadows of the polluting
coke works, the most unpopular landmark in the borough had
simply vanished. Years of articles published, letters sent, surveys
delivered, and petitions signed had prevailed. The pen was indeed
mightier than the sword. The Monkton dragon was finally slain.

In the immediate aftermath, a debate ensued around what the
site would be used for in the future.[252] Many bids were made by
planners to the Department of the Environment, in a necessarily
lengthy process.[253] Within a year of the final demolition, the site
had become a focal point for vandals, with thieves looking to steal
the remaining generators,[254] and an arsonist setting two filtration
tanks ablaze that were perilously close to ten thousand gallons of
oil.[255] A major fire disaster was averted thanks to the speed of the
emergency services, but the event expedited the calls to clarify
the site's future.[256] A new housing estate was considered by the

town's development committee.[257] A leisure centre with a golf course was proposed by the council. Many residents wanted an extension to the city's light rail service. Jennie felt strongly that the site should become Green Belt land, at least in the short term.

> The people of Hebburn want a bit of space in the town. In the past, every piece of available land has been built on. It's important to have green areas for future generations.[258]

— Jennie Shearan.

The council ultimately agreed with Jennie and decided not to earmark the area for any form of industrial or housing development in the short term.

Jennie, provident as always, looked to the future handling of the pollutants that remained on the derelict land.[259] When the site was finally cleared, there were six decades of hazardous waste that had accumulated and needed to be dealt with. The responsibility for the remediation strategy fell to English Partners, a government agency dedicated to redeveloping urban land. With the exact nature of pollutants on the site still unknown, ERM Enviro Clean was appointed in January 1994 to undertake an assessment of the ground contamination. The investigation was carried out over a three-month period and identified a number of severe chemical hazards, particularly on the southern part of the site. The wounds that the coke works had left behind were manifold. Polycyclic aromatic hydrocarbons were found, which have been linked to stomach and bladder cancer. Also present was spent oxide, which is a reddish-brown powder that is formed when coal gas is purified, and inhalation of its fumes can cause a fever and chest tightness. Also detected were phenols, which are acidic toxic white crystalline solids obtained from coal tar that are highly irritating to the skin and eyes. Other derivatives from the viscous tar, such as the irritant xylene and the flammable benzene, were found on the site.[260]

Geo-technical analysis was carried out to investigate the interaction of the soil with man-made materials. The research confirmed that the contaminated material would have to be removed prior to any development. The experts reported that any future housing development would necessitate a comprehensive reclamation scheme.[261] However, the £9 million price tag for leaving the site completely clean of any contamination and buried structures proved cost prohibitive.[262]

With residential housing ruled out, the focus switched to the issue of how to handle the poisonous sludge that persisted there. Clean-up options were discussed throughout 1995 between South Tyneside Council and the developer. The preferred method that the regeneration authorities were considering was engineered cells burial. The health risk of this type of reclamation method is that the plastic capsules can weaken and burst over a 20 to 40 year period. Jennie challenged the planning officers to question the council's decision-making process. She wrote to the Government Office for the North East expressing her concerns that the waste would not be removed to a treatment plant, but instead remain only yards away from people's houses.

I fought long and hard for the site to be cleaned up for the people of Hebburn and Jarrow and I cannot rest until I know for sure that it is safe for future generations. I am not necessarily concerned for my own health. I am in my seventies and there is probably not much likelihood of any such catastrophe occurring in my lifetime anyway. What I am worried about is down the track, if those capsules will leak. You cannot deny that such an event would not risk human health once the pollutants are in the ground. They can find their way into the water consumed by local people.[263]

— Jennie Shearan.

Ultimately, the cheaper engineered cells burial method was chosen. The toxic ground was excavated and placed in plastic-lined capsules, with the encapsulated material then transported to the northern part of the site for burial. Jennie requested all information relating to the buried capsule to be put in the public domain, and regular tests to monitor its safety. The concern Jennie showed for the health of future residents is a testament to her continuously forward-looking empathy. Today, 30 years after Jennie first expressed her apprehension, the Environmental Sustainability Service at South Tyneside Council has confirmed that the area does not show any poor vegetation growth, indicating that the engineered cells are still intact.

Later that year, the Inspectorate of Pollution was dissolved, and the Environment Agency was created. The Monkton Coke Works case had contributed to official recognition of the profound impact of pollution on the public, and the need for a permanent agency to monitor, regulate, and limit pollution of all types, including that of major industries and the waste they produce. Alas, this was a pyrrhic victory. The government had changed the packaging but the product remained the same: strict enforcement of legislation and the financial resources to support it were not forthcoming. Every government since Thatcher's leadership has continued to squeeze public spending,[264] resulting in the downsides of big business going unchecked. In later years, Wendy Le-Las interviewed for a role in the Planning Inspectorate. As is customary at the end of an interview, Wendy was asked if she had any questions. 'Yes', she said, 'do I always have to accept what the Environment Agency tells me?' The reply was no, provided there was evidence to the contrary. Thus, the onus still remains on the community to disprove the case put forward by polluting firms. Today, these firms are mostly registered in tax havens, thus they pay no tax to the British Government and have significant resources at their disposal to crush community dissent.

During the years immediately after Monkton Coke Works closed down, a number of former employees who had worked at the site between the 1940s and 1980s began suffering from cancer, emphysema, asthma, and chronic bronchitis. It became apparent that the Inspectorate of Pollution's declaration in 1990 that workers exposed to the plants' emissions would not suffer adverse effects, was incorrect. The affected workers took their case to court and were represented by a specialist workplace illness lawyer, who articulated their litigation claims.

Hundreds of former coke oven workers are now suffering from terrible conditions, simply because of the work they carried out on a day-to-day basis. Employees have a basic right to be able to go to work and return home safely at the end of the day.[265]

— Bill Sandham.

The price of employment had been premature death. Arguing that British Coal failed to both correctly assess the risk of working with coke ovens and adequately protect workers from harmful dust and fumes, the former employees got their fight for justice heard. Many of the former workers with lung cancer became entitled to industrial injuries disablement benefit, subject to certain employment-related criteria. For the section of the workforce who only had seen Jennie's campaign as a threat to their economic well-being, they could now understand her concerns about the health of her neighbours who had also been exposed to the same noxious gases on a daily basis.

What of compensation for the residents of Monkton Lane Estate? Rather than working at the plant for eight hours a day, five days a week, the residents had lived beneath the poisonous chimneys of Monkton Coke Works for 24 hours a day, 365 days a year.[266] Countless residents continued to suffer from serious medical issues, including debilitating respiratory ailments. Sadly, there was no legal aid available, and no progress was made on a large number of personal injury claims from the local population whose lives had been affected by Monkton Coke Works.[267]

By 1995, a decision had still not been made about the future use of the site. The contents of a Unitary Development Plan that was subsequently produced by South Tyneside Metropolitan Borough Council were very unsettling for Jennie and her fellow action group members. The proposal laid out a vision for a hypothetical heavy industry manufacturing plant that would employ over 4,000 people. For an area in such need of employment since the decline of the mining and shipbuilding industries, it was considered by the local Labour Party as an excellent use of the site.

For Jennie and the action group, however, the prospect of turning over 200 acres of Green Belt land back to heavy industry represented an unravelling of decades of campaigning for the polluting site to be cleaned up. They, of course, did not object to job creation but wanted to first verify that the land was safe to be used, and then understand whether a different type of industry could be established there to provide environmentally friendly employment for the area.[268] The local journalists who had closely followed the action group's campaign to clean up Monkton Coke Works naturally sought Jennie's opinion on the matter.

With a front-page headline shouting 'What price jobs?', a subheading stating, 'Anger at plan to create work for up to 4,500', and a photo of Jennie standing next to a fellow action group member at the former site of Monkton Coke Works, the pair were crassly depicted as obstinately objecting to employment in the area. It was to be the beginning of an inaccurate narrative that was propagated by the *Gazette* for many months.[269]

Jennie decided, as was her wont, to address the issue head-on. She visited the newspaper's headquarters to directly address the issue with the senior editor.[270] Jennie was told that the newspaper acknowledged that the story had been sensationalist, and would write an apology piece. In the event, the apology piece never materialised, just like the fictitious manufacturing plant and its fabricated number of jobs.

Despite the lazy journalism attempting to create a false impression about Jennie and her views on utilising the site for employment, the town of Hebburn knew of Jennie's rectitude and continued to recognise what she had done for them. In an open letter to the *Gazette*, local resident John Badger wrote the following:

> I suggest that you have missed the point with your article.
>
> There can be no question that the borough needs jobs. Real jobs and plenty of them. But the manner in which your article is worded creates the impression that a 'Nissan-style developer' is just over the horizon. Where is the evidence for that?
>
> As for the claim that 'people living near the coke works want the land retained as Green Belt, not for industry', then little wonder after all that they had to put up with for years, and are still living the legacy of that time.
>
> What I consider to be most unfair is the portrayal of Mrs Jennie Shearan as the villain of the piece, as someone who is hell-bent on depriving people of much-needed jobs at all costs. Those of us who know Jennie, know otherwise.

No. Jennie Shearan is blameless. The whole idea
has been born out of a desire to further certain people's political careers and has nothing to do with
creating jobs.[271]

Five years after the plant had been demolished, Jennie, now 75
years old, was walking towards Finchale Road bus stop to go to
the New Town. A similarly aged woman approached her and
said, 'I know your face, and your voice. You're the woman who
gave us clean air'. Clean air is what Jennie had always wanted
for the residents of Monkton Lane Estate. It was a poignant
moment for her and a reminder of the impact that the Hebburn
Residents' Action Group had made on the community.

Her successful campaign had laid the groundwork for the site
to metamorphose from a polluting industrial plant into a space
of which the town of Hebburn could be proud. The question
remained what this site would be used for. Once all the debris
had been removed, there was a twenty hectare space that was
roughly trapezoidal in shape, with the diagonal line of the former
Bowes Railway line continuing to divide the site in two.

In 1998, the land where Monkton Coke Works once stood
was acquired by One North East, the development agency for
the region. The agency received monies from the Homes and
Communities Agency's National Coalfield Programme to restore
the site. On the strict understanding that the land could never
be used for residential purposes, it was decided that half the land
would be designated as public open space, with the remaining
ten hectares to be redeveloped as a business park.

The Hebburn Residents' Action Group were among the members of the community who were invited to contribute to the vision for how the public space should be developed. Jennie's focus remained on the youth of Hebburn. In a letter to the creative agency that was hired to oversee the project, she wrote the following:

> We've lived through hell and now we want a better environment for people. It is important for the next generation to have a better start; we want our children to be able to run and breathe clean air.[272]

— Jennie Shearan.

As part of the toxic excavation process, excess ground from the southern part of the site was formed into mounds to give variety to the once flat land.[273] Clean subsoil was imported to the northern part of the site to establish woodland in this area. Community involvement was encouraged to instil a feeling of pride and ownership in the local environment. In collaboration with Groundwork South Tyneside and Newcastle, over 30,000 trees, wildflowers, bulbs, and shrubs were planted, many by the local schoolchildren. The foxglove (*Digitalis purpurea*) and red campion (*Silene dioica*) plant species were specifically chosen because they attract butterflies and bees, which in turn help the flowers to spread and thrive.

By 1999, the site, renamed Monkton Community Woodland, was nearly ready to be opened to the public. It had taken eight years for the clean-up process on the contaminated land to be complete. It was now a home for nature, spanning four and half kilometres of tracks, with four woodland paths.[274] The new footpaths and cycleways across the reclaimed space were created to provide several vital links between communities and a network of green spaces and wildlife, as well as opportunities for volunteering and regular health activities. The northern section of the woodland had been extended to form part of the Great North Forest. The former Bowes Railway line, remaining intact through the site, was now part of a long-distance cycleway linking to other green spaces both east and west of South Tyneside. Now the young cross-country runners of nearby St Joseph's school finally had somewhere safe to do their training!

Next door to St Joseph's school on Mill Lane sat South Tyneside College, just a few hundred metres from the former coke works site. This was an education facility for the offshore and ship repair industries and was set up to undertake large-scale pressings, rolling, and welding. It delivered a wide range of tailor-made courses and was the largest supplier of engineering apprentices in the region. The college had recently instituted an artist-in-residence position, intended for a local with knowledge of metalworking processes, and a deep affiliation with the area. A feasibility study was launched at the beginning of 2000 into the creation of public artwork for the site.[275] When the resounding conclusion was positive, South Tyneside College's artist-in-residence William Pym was chosen to head up the project. The goal of the initiative was to enhance the woodlands by providing a focal point for the location, and foster community support for the scheme. Following a £125,000 grant from One North East, Pym and his fellow welders began developing a sculptural piece that reflected both the historic and future use of the site. Given that steel is produced with the help of coke, it was agreed to be the most appropriate material for the project.[276]

Across March and April 2000, numerous sections of the Hebburn community had input into the artwork. Local schools, youth workers, former employees at the coke works, a local history group, and the action group were all consulted. Jennie sat on the steering committee, alongside school children from St Joseph's. They shared memories about the plant and their aspirations for the development of the site. It was decided that a swarm of bees would be a fitting metaphor to incorporate the many thoughts and feelings on the issue. With Pym supervising the project, his fellow welding students helped with the steelwork to bring his vision to life.

> The bees are like the image of the industry that's leaving and the notion of bees being good and bad. They sting, but they are also productive in the same way the old industries were very productive. I like to think my work is optimistic. What affects me is the extraordinary stories of people and their communities that have something unique. It's pulling that out and expressing that, because often what is true for lots of people is true really powerfully for the individual.[277]
>
> — William Pym.

With the art pieces completed and installed, Monkton Community Woodland and Business Park was officially opened in June

2000. Rising from the ashes of this former notorious blot on South Tyneside's landscape, *The Swarm* and *The Hive* sculptures were an impressive visual statement on the transformation of an industrial site that was once such an eyesore. These innovative pieces graced the skyline, symbols of the natural world looking towards a more hopeful future. Fluttering among the artwork and trees, large skipper and speckled wood butterflies began to populate the area, with bullfinches and willow warblers freely joining them, while rabbits and squirrels jumped among the daisies and lilacs. It was a profound rebirth that was a world away from Monkton Coke Works.

The launch of the site was officiated by Parliamentary Minister Hilary Armstrong, who paid tribute to Jennie in her speech.

> Jennie is one of those women in the North East who was determined to show what women can do. All her life she has shown concern for her local community, and it is women like Jennie who have made it easier for women like me to do what I do.[278]
>
> — Hilary Armstrong.

Jennie, now 77, was also asked to give a speech at the event.

> Today is a wonderful day for the people of this area. Where once stood Monkton Coke Works and black clouds of pollution now stands the beginning of a pleasant environment. We fought long and hard for the right to breathe clean air.[279]

— Excerpt from Jennie Shearan's speech.

Today, in 2022, Monkton Business Park houses twenty companies, with trees adorning the main entranceway. One corporation alone employs in excess of 250 workers, a far greater number than Monkton Coke Works ever did at any point in its history. Among the firms with offices at Monkton Business Park are multinational, blue-chip, construction company, Hitachi; sustainable infrastructure engineers, Kier; and specialist recruiters for the engineering and construction industries, CDM Recruitment.[280] Thus, Jennie and the action group contributed to getting clean air for their community, without sacrificing much-needed employment in the area. Not only did the number of job opportunities from the site significantly increase, but the types of companies that are operating on the site are continuing the rich tradition of employing Tyneside people involved in heavy industry, but now in a clean way.

Beyond a natural green environment that today employs hundreds of people, Jennie left her mark in another way. The Monkton Coke Works case was the catalyst that led to the setting up of the Environmental Law Foundation in 1992. Soon after the case concluded, several leading barristers, solicitors and environmentalists, including Dr Wendy Le-Las and Charlie Pugh, convened to address the need for an organisation to help use the law to protect the environment. The Environmental Law Foundation was born, a charity that aims to empower local people to address all manner of adverse environmental issues and challenges, and get them access to justice. Many communities like Monkton Lane Estate exist across the UK, whereby ordinary

people struggle to be heard on matters affecting the environment in which they live. They are often disadvantaged due to a
lack of resources or information to get their voices heard. The
Environmental Law Foundation was established to promote
collective, good decision-making which is at the very heart of
civilised, democratic, and stable societies. Thirty years on, the
charity today deals with over 200 cases annually, with over 150
lawyers giving up their time to provide pro bono legal assistance
to the value of as much as £1 million a year.

Jennie's campaigning not only improved the quality of air
that Hebburn breathed, leaving behind a woodland with open
spaces and clean employment for her community, but also helped
make this dream a reality for hundreds of towns across the UK
for future generations.

Jennie about to meet with the Inspectorate of Pollution

SOUTH TYNESIDE Times

Vital help for coke works campaign

LEGAL AID LIFELINE

HELP looks close at hand for Hebburn residents needing legal aid in their battle against pollution from Monkton coke works.

An action group campaigning for a clean-up at the works came to a standstill when their lawyer Ian Bryson, who offered his services free, moved to London.

Their fight hit a brick wall when a bid at going solitary at South Tyneside Council when the Pollution Inspectorate to the Government in full of confident satisfaction was won.

But former Tyne and Wear County Councillor Jenny Shearer, a member of the action group said it would enable the community to afford a far more aggressive public inquiry on Tuesday, January 30.

"Just was marvellous," said Mrs Shearer. "Just to get a job in London and although we still would have adjourned, he obviously can't be involved as such.

"British Coal have it also close began working for them who will represent them at the inquiry, so to have any chance at all we need to interpret on an equal footing.

"But on Monday I had a phone call from a woman in Canterbury who represents a consortium of barristers."

Mrs Shearer added that the woman was very keen to get to the fight and optimistic about finding a law who willing to help.

"I sensed that it would bear on her face," added the ex-council chairman, "because obviously we can't compete on a financial level.

"Not she phoned on back shortly afterwards and said normally this people even knowned out to London and over to Durham."

The candidate say the works winter gives out hundreds of filth spews, causing their complexion and clogging in clothes and buildings.

Action group leader Jenny Shearer outside the controversial Monkton coke works. (Ref. K207331)

Action group leader Jennie slams Michael Heseltine's action

Plant decision is condemned

By TERRY KELLY

ACTION group leader Jennie Shearan today condemned Environment Secretary Michael Heseltine for overturning a High Court decision on a power generation and chimney scheme at Monkton Cokeworks.

Protesting residents and South Tyneside Council officials were shocked at yesterday's announcement, which came after two costly public inquiries into the scheme.

The decision could give the green light to a major gas turbine power station and 160ft chimney at the plant.

But cokeworks owners Coal Products Ltd say they have no immediate plans to resume production at the mothballed plant.

Protesting residents living near the controversial Hebburn works were delighted when it was mothballed before last Christmas because of commercial reasons, due to a downturn in the market, after years of protests about pollution from the works.

But Mrs Jennie Shearan, leader of Hebburn Residents' Action Group, who led the fight for a clean-up of the works, said the decision is a blow - but vowed residents will fight on.

She said: "We have had a few months of heaven, and we don't want to go back to hell. We will fight this decision again in the High Court, if necessary. It was the residents who took our case to the European Parliament - not the council - and we will do it again."

Mrs Shearan is now in contact with the action group's barrister, Mr Charles Pugh, to determine what their next step should be.

Mrs Shearan, who lives in Melrose Avenue, only yards from the works, said she was "disgusted" by the Department of the Environment's decision, which she believes was taken by a junior Government official not Mr Heseltine.

She said: "I don't believe Mr Heseltine has been in the job long enough to have given read the papers. He is supposed to be for the environment, but has willy-nilly given this decision without a thought."

Mr Heseltine's department has instructed Her Majesty's Inspectorate of Pollution to carry out a full appraisal of the effects of the proposed plant and suggest whatever pollution controls are thought necessary.

Nothing

A spokesman for Coal Products said: "We have heard nothing official from the DoE.

"But judging from the stories we have heard through the media, it does seem a satisfactory outcome to our appeal.

"But we have no immediate plans to resume production, as the plant was mothballed for commercial reasons."

MONKTON at night during full production at the cokeworks.

A VICTORY AT MONKTON?

HELL-HOLE!
Cokeworks blasted over Thursday's emission...

FLASHBACK ... a half-victory for the Monkton campaigners when CPL announced a £500,000 anti-pollution scheme. Yesterday Michael Heseltine said no immediate waste precautions were needed.

Cokeworks crusader's long battle

By EDDIE BROWN

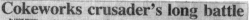

● Mrs Shearan at her home with the cokeworks in the background.
Picture: TOMMY MACKIN

No coke
... without fire!

AFTER more than a year of uncertainty, it was finally announced this week that Monkton Cokeworks is to be demolished.

Residents were jubilant when the works was mothballed just over a year ago, after an anti-pollution campaign that hit the headlines for more than five years.

Now, British Coal chairman Neil Clarke has revealed that a further downturn in the export market for the coke produced at the plant has made demolition inevitable.

An ecstatic Mrs Jennie Shearan, who has led the Hebburn Residents' Action Group in their fight for cleaner air around the works, said the announcement is welcome news for everyone who, she claims, has suffered as the result of pollution from the works.

There had been rumours and counter-rumours ever since the works was mothballed in late 1990, with the loss of more than 200 jobs. Despite the decision, residents would not believe the plant was finally doomed until it was officially announced that it was to be demolished.

This news eventually came on Thursday of this week, and Mrs Shearan said the children around the works will now be able to breathe clean air.

As for the future of the site, the first priority will be to remove the various pollutants which have accumulated since the works went into production in the early 1930s.

Then, it was a largely green field site, far away from any housing. But with the 1950s came large-scale council house building. In 1964, this mushroom growth continued when the Lukes Lane Estate was built and hundreds of residents found

their homes overlooked the plant, whose fumes, smoke and familiar flame became something of a local landmark.

But Monkton Cokeworks became possibly the most unpopular landmark in the borough, especially when production of coke at the plant was stepped up in the last

⇧ ON THE MARCH ... protestors, including Jennie Shearan, stage a demonstration outside the works. Their ceaseless efforts have finally been rewarded.

☐ IN the light of the decision to demolish Monkton Cokeworks Gazette district reporter TERRY KELLY looks back on the bitter battle by residents to rid themselves of an eyesore.

15 years.

The protest gathered momentum in the late 1980s, as Mrs Shearan, whose council house in Melrose Avenue stands only a stone's throw from the plant, decided enough was enough and gathered her action group around her to take on the might of British Coal and Coal Products Limited.

This protest snowballed, and soon the campaign was ultimately taken to the European Parliament at Strasbourg.

There was also a health survey carried out by the action group before an official £50,000 health inquiry was launched by experts from Newcastle University.

As the protest reached its height, it was announced the plant was to be mothballed, officially because of a downturn in the foreign export market for

the coke produced at Monkton.

But Mrs Shearan and her fellow protestors are convinced that British Coal and Coal Products Limited were worried about the results of the health survey into the effects of pollution from the plant. If we accept this argument, it has been peo-

ple power which forced the closure of the works.

None of the residents wanted to throw 200 men on the dole, but the tide of public opinion was so strong against the works that closure - in retrospect - seemed inevitable.

The long protest campaign has seen residents pitting themselves against a massive corporation - and winning.

There seemed to be little communication between both sides, with British Coal apparently unwilling to recognise the real anger of local people at heated public meetings over development plans at the cokeworks.

Fortunately for residents, their campaign coincided with the rise to prominence of Green issues and their fight was taken up by the media.

Jarrow MP Don Dixon was understandably concerned about the job losses at Monkton, but this week welcomed the

demolition decision.

He would like to see Central Government inject capital into a scheme to provide industrial units on the site. This scheme is already being supported by South Tyneside Council, who say the land would be ideal for such a development.

As for the residents, they are simply glad that they can breathe clean air

again, after decades of living next door to one of the worst environmental eyesores on Tyneside.

But the numerous compensation cases that the action group are pressing for people who claim their health was damaged by pollution from the plant illustrate that the legacy of Monkton Cokeworks will remain for years to come.

PAUL KELLY on the health problems caused by Monkton plant

Chokeworks!

POLLUTION from notorious Monkton Cokeworks damaged the health of local people, a £60,000 health survey has found.

Residents living on estates around the mothballed Hebburn plant were more likely to suffer respiratory problems than people in other parts of South Tyneside, the Newcastle University survey reveals.

However, research shows people living nearer the plant had mortality and cancer rates not significantly different to those in other areas of the borough.

The three doctors who carried out the two-year survey, Raj Bhopal, Suzanne Moffatt and Peter Phillimore, revealed their findings to councillors yesterday and at two pubs in meetings in Hebburn.

Dr Moffatt said: "The level of respiratory ill-health in the populations living close to the Monkton Coking Works was worse than expected.

"The evidence suggests much of the excess of respiratory ill-health is likely to have arisen as a result of exposure to emissions from the coking works."

Suffered

Questionnaires sent to hundreds of homes near the works, including nearby Lukes Lane estate and Jarrow's Primrose estate, showed adults, and particularly children, suffered more from sinus problems, hay fever, chronic bronchitis and allergies, than was the norm in South Tyneside.

The research team did not find any evidence of lower birthrate, abnormal birth ratios or a greater proportion of stillbirths.

Doctors found that respiratory problems did not diminish when production at the plant stopped due to the miners' strike of 1984 or after the final closure in October 1990.

But examination of GP records did reveal a higher level of headaches, breathlessness, coughs and phlegm, during the times production was at its highest.

Residents questioned expressed an expectation the plant's closure would lead to a small improvement in their health.

Many more said pollution-related problems were causes of stress and anxiety to them.

When the cokeworks was built in 1937 it occupied a green field site but by 1967 it was surrounded by housing estates to the North, East and South East.

Residents became concerned about both the nuisance and health effects of living near an industrial plant which emitted airborne pollutants.

In response to a lengthy and sometimes acrimonious debate between the residents and the management of the coking works, the borough council funded the new study.

Jennie Shearan, who led a clean-up campaign at the plant, said the research did not prove pollution from it had not led to residents contracting cancer.

She added: "I know that it took 20 to 30 years for research to prove people died from asbestosis. South Tyneside Council allowed production at the plant to double. How long will it take until the effects that decision had on health will be known?"

Hebburn councillor Neil Rooney said: "Hopefully this report will be comforting to the people of the area. Many have been needlessly suffering stress worrying about possible damage to their health. If this research puts their minds at ease the £60,000 will have been money well spent."

The cokeworks is to be demolished later in the year.

PROTEST LEADER ... Jennie Shearan.

RESEARCH LEADER ... Dr Raj Bhopal.

TWO-YEAR RESEARCH ... Dr Suzanne Moffatt.

END OF AN ERA ... Monkton Cokeworks is to be demolished later this year.

The final demolition of Monkton Coke Works

Jennie at the opening of Monkton Community Woodland
and Business Park

Art produced by local artist William Pym at Monkton
Community Woodland

Monkton Community Woodland today

The difference in view from 1 Melrose Avenue

EPILOGUE

In 1987, the United Nations World Commission for Environment and Development published a report called 'Our Common Future'. The General Assembly had agreed on the urgent need to define the measures required to deal with the problems of protecting and enhancing the environment. The resultant paper outlined a mandate for change; that the environment could not remain a side issue in central decision-making when striving to meet humanity's goals. Referencing numerous examples of nations retreating from social concerns in pursuit of economic gain, 'Our Common Future' called for coordinated political action and responsibility. The report went into great detail about the unprecedented pressures on the planet's lands, waters, and forests, and the concomitant environmental degradation. It called for a new era of economic growth that was socially and environmentally sustainable, strongly stating the belief that people could build a more prosperous, more just, and more secure future. In doing so, it introduced the principle of sustainable development, defined as 'development that meets the needs of the present without compromising the ability of future generations to meet their own needs'. The report did not offer a detailed blueprint for action, but instead a pathway by which the peoples of the world could enlarge their spheres of cooperation.

Today, the report is regarded as a milestone in triggering international awareness and discourse on the importance of global sustainable development.

> When the terms of reference of our Commission were originally being discussed in 1982, there were those who wanted its considerations to be limited to 'environmental issues' only. This would have been a grave mistake. The environment does not exist as a sphere separate from human actions, ambitions, and needs, and attempts to defend it in isolation from human concerns have given the very word 'environment' a connotation of naivety in some political circles. The word 'development' has also been narrowed by some into a very limited focus, along the lines of 'what poor nations should do to become richer' and thus again is automatically dismissed by many in the international arena as being a concern of specialists, of those involved in questions of 'development assistance'.

> But the 'environment' is where we all live; and 'development' is what we all do in attempting to improve our lot within that abode. The two are inseparable. Further, development issues must be seen as crucial by the political leaders who feel that their countries have reached a plateau towards which other

nations must strive. Many of the development paths of the industrialised nations are clearly unsustainable. And the development decisions of these countries, because of their great economic and political power, will have a profound effect upon the ability of all peoples to sustain human progress for generations to come.[281]

— Our Common Future.

In the same year that 'Our Common Future' was published, the Hebburn Residents' Action Group was formed. The tireless activism of the action group closely mirrored the most salient issues that 'Our Common Future' outlined. Even though they had limited resources, the campaigners organised themselves and took responsibility to ensure sustainable human progress. The decisive action they took led to front-page headlines and TV documentaries that highlighted the urgent need to both protect and enhance their environment. They overcame multifarious obstacles in their attempts to make the environment, not economic imperatives, the central issue in decision-making. They recognised the interlocking matters of economic development and environmental protection and advanced the concept of sustainable development. Not only that, but they contributed to a solution from which future generations would benefit. If Monkton Coke Works was an example of how 'Our Common Future' highlighted where economic growth had been prioritised ahead of the environment, then the efforts of the Hebburn Residents'

Action Group were a microcosm of the global vision to which 'Our Common Future' aspired.

The residents of Monkton Lane Estate were undeniably the victims of this story. However, the managers of Monkton Coke Works were also victims, and moreover, victims of the same perpetrators: negligent town planners.[282] The purpose of the plant was environmentally worthy; to create smokeless coke. Unfortunately, Hebburn Urban District Council did not fulfil its role in mitigating risk to the public. When the councillors gave planning permission to build social housing so close to a coke works, they did not take into consideration the effect that it would have on the future residents who had to live there. It may have been caused by inadequate communication or intra-agency conflicts or, as Christopher Miller posits in his study on planning and pollution, perhaps the planners were single-mindedly focused on 'maximising total utility'[283] of all available land. Whatever the cause, the effect was that the council gave only peripheral regard to the environment, with their priorities very much focused on the health of businesses and jobs. The Local Planning Authority allowed the construction of not one, but two, housing estates on the doorstep of Monkton Coke Works. By its very nature, a polluting coking plant is impossible to clean up. The fundamental problem that both the residents and the facility managers encountered was that irresponsible planners had allowed these two entities to co-exist. With no means for British Coal to seek redress in a negligence action against the Local Planning Authority, the firm was left to fight it out against the Hebburn Residents' Action Group.[284]

Once Jennie's mission was finally achieved, a local journalist made this observation:

> Jennie and the other campaigners deserve an award for their tenacious decades-long struggle against the fumes. But you don't generally get prizes for taking on the establishment.[285]

I strongly agree with his sentiments about the need for recognition of their endeavours, and I hope that this book has served as a felicitous memorial of their unswerving will and bravery.

Thanks to Jennie and the action group's cumulative efforts, their fervent campaigning eventually triumphed over the dreadful hazardous by-products of coal being transformed into coke. Together, the activists replaced alienation with solidarity, muck with cleanliness, and helplessness with hope.

The vision, courage, and staying power of the green crusaders were exceptional, and even more admirable given the gender roles that they had to fulfil and the gender discrimination that they had to endure. They all experienced 'time poverty'[286] as they balanced the burden of household and caring responsibilities with their environmental activities. Moreover, they were fighting against a system that had institutional sexism and they had to cope with intimidation from some men in the local area who understandably had more of an interest in their own family's economic well-being than in environmental justice for the community. There's an old saying that if you want to go fast, go alone, but if you want to go far, go together. By listening, engaging, and working together, the action group tackled a seemingly hopeless issue. Together, they valiantly gave their community a seat at the table and their collaboration helped power the movement towards meaningful change.

It was Jennie, the intrepid leader and public face of the action group, always on the frontline, that carried the heaviest weight. By bringing together fellow housewives to form the Hebburn Residents' Action Group, she empowered similar women to believe that they could be agents of change in their community. All the while, she juggled the challenges and anxiety that came with caring for her unwell husband who she loved dearly. From a young age, she had believed in social, economic, and political equality of the sexes. As a middle-aged woman entering into the male-dominated world of politics, Jennie emboldened generations of women, young and old, to stand tall. She was someone who had to fight to be heard in order to deliver material change for her town as councillor, and she took that same irrepressible spirit into her battle against the male-dominated government departments to clean up the coke works. In seeking to constantly represent, support, and promote women wherever she could, from her early political career to her leadership of the action group, she played a role in breaking down the gender stereotypes that had existed throughout her life.

Jennie was fearless, overcame innumerable obstacles, and never, ever gave up. She was never afraid to passionately share her perspective and she built strong relationships across the country because of her authenticity. Alongside her disarming honesty was a boundless determination. She encountered numerous setbacks throughout her fight for the rights of her town but was resilient in exploring all possible solutions and overcoming adversity. Jennie knew that failure was not a permanent condition, and every time she got knocked down, she just got back up again. Her unshakeable values, strength of character, and formidable spirit conquered these countless challenges. As the leader of the action group, she entered into a legal arena in which she had no formal education. She did not shy away from tough conversations in the courtroom or candid information-gathering in the streets. Instead, she and her fellow activists diligently unearthed the human and environmental impact of a negligent big energy firm that prioritised profits over people.

We all just have one life on this planet. A country's economy and politics are managed with divergent agendas that can create complex situations that do not have a simple resolution. However, sometimes there is simply a right thing to do, especially when it relates to many lives. And sometimes, it takes an incredible amount of humility and persistence to get things put right, especially if it is not the easiest or most popular way forward. Jennie altruistically strived for her goal of clean air day in, day out, for years. It was a marathon that required stamina and perseverance. Regardless of the hardships she had to endure, she kept on fighting, because she considered it a moral imperative to make a change. Nothing would deter her in her quest to end the turmoil for the people of south Hebburn. It was Jennie's grit and drive that made her dream a reality. She achieved so much, with so little, doing what so few were willing to do, with extremely limited funds and none of the modern technologies that make communicating and accessing information so easy today. Her consistent commitment to equity for her community was unwavering at every stage, from the urgent and decisive lobbying of governing bodies that she and the action group undertook, to her tireless work to ensure the clean-up was handled correctly, long after the final demolition took place.

Moreover, Jennie operated with admirable vision and foresight, both in how she led the campaign and how she saw it through to the very end. Being so committed to doing the right thing for the people of Hebburn, she knew that her approach had to be comprehensive if they were to be taken seriously in court and in the media. When the moment arrived to represent her town's plight in public, Jennie was meticulous in her planning and delivered her beliefs passionately.[287] In many ways, she was ahead of her time. She understood the value of grabbing people's attention, and deployed countless tactics and tools to lay bare the environmental inequity that her community faced. Jennie made her green campaign go viral decades before millennial marketing managers had even coined the phrase and she forced a dialogue about the coke works' 'carbon footprint' before the expression existed.

Jennie intimately understood what employment meant to this proud community. Throughout her life, she had been immersed in the world of heavy industry and was a member of the working class who relied on it for their socioeconomic well-being. Jennie fought tirelessly to grant her beloved community both clean air and clean employment. Ultimately, her redoubtable efforts left a deeply positive and long-lasting impact on her town, giving thousands of people hope for a stronger and healthier future and transforming their quality of life.

She was a prodigious force for good, whose every action revealed a magnetic environmentalist who cared deeply for her community. By seeing the threat that Monkton Coke Works posed to human lives in her community and striving to create the type of world that she felt the residents should live in, Jennie elevated consciousness of a major environmental problem. In so doing, Jennie became a poster child for the green movement in the region, at a time when green issues were barely gaining prominence on the political agenda. She helped make environmentalism a newsworthy topic at a national level. During the time that Jennie was leading the long battle for the clean-up of Monkton Coke Works, environmental activism was not as widespread and mainstream as it currently is in 2022. Today there is a sense of urgency around protecting the environment and raising awareness of the human costs of heavy industry, but back in 1980s North East England, a general understanding of these matters, and acceptance that this issue was a priority, was far lower. Jennie got the topic of environmentalism onto the front page of newspapers and onto national television programmes.

Jennie's gallant efforts in championing the human right to access clean air also catalysed the creation of a national charity that today helps some of the most disadvantaged communities in the country that are seeking environmental justice. Her eco-movement highlighted systemic and long-standing political ecology issues that still exist across the country today. Her actions to protect a populace that was disproportionately bearing the cost of environmental harm laid the groundwork for the Environmental Law Foundation which continues to help communities throughout the UK.

As I sit here writing these final words, I am astounded at just how far-reaching the remarkable legacy is that Jennie left behind.

Jennie was my grandmother. I am the son of her youngest child, Maria.

I was born in 1983 and grew up largely oblivious to my grandmother's campaigning. She was a constant throughout my childhood. My twin sister and I would often stay with my grandmother on weekends. We were ten years old when the last chimneys were finally demolished. Being so young at the time of her battle with Monkton Coke Works, my only recollections of her in this time period are of the many happy memories that we shared, receiving countless hugs and kisses, dancing in her living room to ABBA, and enjoying her delicious rice pudding.

As I grew older and progressed through school, she would encourage and advise me. We developed a special bond and her words had a tremendous influence on me as I navigated my teenage years. She always had a motto at the ready to help me confront any challenge I was faced with. Learning about what my grandmother had to go through to win her battle for the people of Monkton Lane Estate to be heard, I can now appreciate where these phrases were conceived and where they were put into practice, time and again.

'Always look people in the eye, Gian. You're no better or worse than anyone.' She was always true to herself, and had an innate sense of fairness and a need to be honest with everybody.

'Shy bairns get nowt, Gian.' This is a Geordie proverb that translates as 'shy children get nothing', by which she meant that you should not be afraid to be assertive and ask for something, or put yourself forward if it is for something you believe in. She always listened to her heart and dreamed big, genuinely believing that people could achieve anything as long as they did not give up.

'You never stop learning, Gian.' Throughout her life, she was open-minded, curious, and courageous. She asked questions and took chances, treating every new challenge as a positive learning experience.

'You can only do your best, Gian.' My grandmother knew that there is a limit to what one person can achieve, but she also realised that the limit was so much further away than many of us are led to believe. She embraced that journey. As a result, she never cut corners in whatever she did, always giving it her absolute best shot.

The statement that she would often make in her gentle Geordie accent which endures with me was 'I'm a rebel, Gian'. It was only once I wrote this book that I fully understood what she meant!

In 2002, our family celebrated my grandmother turning 80 years old. She was as vibrant as ever, and the vigour in her eyes was undiminished. Surrounded by her sons and daughters and grandchildren who adored her wholeheartedly, it was a beautiful occasion.

Sadly, within a year, she was diagnosed with cancer. She battled the disease stoically, but by the winter of 2004 she was resigned to the inevitability of her predicament. My mother is an avid photographer and has captured key moments as a family on camera throughout my life. With the advent of smartphones, she got comfortable with making short videos and caught a final tender moment between my grandmother and I.

It was just after Christmas, and although the North East has unpredictable weather, the temperature on this particular day was suitable enough for us to venture out to enjoy the crisp air. My father was driving us towards the River Tyne, where we had decided to go for a walk along the quayside. My mother, sitting next to him in the passenger seat, began filming my grandmother and I in the back seat as we sang along to the songs playing on the radio. I had my arm around my grandmother, who was frail but well covered up with a striped scarf and a long wool coat. The radio DJ switched to a new track.

It was Gerry and the Pacemakers. They were a 1960s group that were strongly associated with their hometown of Liverpool, a city where my grandmother had spent some of her formative years. The musicians began playing 'You'll Never Walk Alone'. The song is a quintessential anthem for the people of Liverpool, but the values of sticking together when times get tough, trusting in the abilities of others, and a conviction that better days are ahead, have led to it being embraced across the world. Knowing so little of her past at the time, I could not guess at what my grandmother was contemplating while she softly sang along. I can't help but think now that the uplifting melody and lyrics about solidarity and togetherness took her on a journey through her own lifetime of community service.

> *When you walk through a storm,*
> *Hold your head up high,*
> *And don't be afraid of the dark.*

At the end of a storm,
There's a golden sky,
And the sweet silver song of a lark.

Walk on through the wind,
Walk on through the rain,
Though your dreams be tossed and blown.

Walk on, walk on,
With hope in your heart,
And you'll never walk alone,
You'll never walk alone.

Walk on, walk on,
With hope in your heart,
And you'll never walk alone.
You'll never walk alone.[288]

Shortly before my grandmother passed away, she asked me, 'Will you write a book about the coke works?' She had just entrusted me with the task of transcribing a story that was not only one of the defining moments of her life but also a defining moment for her town. My answer instinctively was, 'Yes', despite having only a very peripheral understanding of what the plant represented and what my grandmother's involvement in its fate had been.

This was the first time that death was a reality for someone so close to me. Moments after we agreed that I would take this on, we said goodbye for the last time. Days later, aged 82 years old and surrounded by her family, who had taken it in turns to be by her side every hour of the day, my grandmother passed away.

That was seventeen years ago. While 1 never forgot about my promise, and I carried my love for her across my journey through life, I felt daunted by the task of truly understanding and documenting what she had achieved. I had always enjoyed reading poetry and writing short stories, but this felt like an insurmountable task. I procrastinated. Months turned into years. I vowed for most New Year's resolutions that this would be the year that I would get this done, to no avail.

As 2021 reached its conclusion, an important milestone was approaching. On 26 November 2022, my grandmother would have been 100 years old. I felt that getting a memoir published by this date would be the most appropriate way to mark her centenary and fulfil my promise.

Ironically, as I dug into the details of this story, piecing together the facts, scouring the carefully labelled archives, and educating myself on obscure terminology and acronyms, I found the project unbelievably fascinating. Delving into the press clippings, recorded television documentaries, and legal documents that my grandmother had left behind, it struck me that she truly had been an unremitting freedom fighter. The piecing together of the chronology of everything that she and the Hebburn Residents' Action Group achieved quickly became a joy.

I hope you have enjoyed learning about their efforts as much as I did.

Jennie's story is an exhilarating and inspirational reminder of the beauty of selflessly fighting for a cause that will benefit generations to come. She was a trailblazer, and when her ambitious vision finally manifested itself, it changed the lives of people in Hebburn and beyond. Jennie is a woman to be celebrated and I am so proud of everything that she accomplished and left behind.

TIMELINE

1922	Jennie Shearan is born
1936	Jarrow March
	Monkton Coke Works is constructed, with a battery of 33 ovens
1952	The Great Smog of London
1953	Jennie moves into newly built housing on Monkton Lane Estate
1954	A new battery of 33 ovens are built, totalling 66 ovens
1981	Jennie secures rent rebates for the residents
1984	The National Union of Mineworkers call a nationwide Miners' Strike
1985	Jennie is elected as Chairman of Tyne and Wear County Council
1987	National Smokeless Fuels request permission for a power station
	Hebburn Residents' Action Group (HRAG) is formed
	HRAG present their petition to South Tyneside Council
	HRAG correspond with Buckingham Palace
	First public inquiry is held
	HRAG stage a peaceful protest at the plant
1988	BBC *Watchdog* documentary airs
	Community charity event at The Victoria Park
	ITV documentary *Bad Neighbours* airs
	HRAG organise their own health survey
	HRAG stage a sit-in at the plant
1990	Second public inquiry is held
	Health study by Newcastle University begins
	Announcement that Monkton Coke Works is to be decommissioned
1991	Michael Heseltine MP overrules the desulphurisation requirement
1992	Announcement that Monkton Coke Works is to be demolished
	Health study by Newcastle University concludes
	The Environmental Law Foundation is founded
1993	Final demolition of Monkton Coke Works
1994	ERM Enviro Clean assess ground contamination on the site
1995	Burial of hazardous waste at the site
1998	One North East acquire the lands
1999	Site is renamed to Monkton Community Woodland
2000	Monkton Community Woodland and Business Park opens
2005	Jennie passes away

ACKNOWLEDGEMENTS

There are many people I would like to thank for helping me complete this work.

Firstly, I am indebted to Jennie Shearan and Barbara Burns. Without the comprehensive and well-organised archives that they both left behind, this project would not have been possible.

I am also sincerely grateful for the valuable input from my family who supported me on so many levels as I pieced this memoir together. Thank you to Maria Rosolia, Antonio Rosolia, Victoria Rosolia, Moira Alexander, Ian Alexander, Brain Shearan, Nicola Alexander-Dent and Neil Alexander. Their anecdotes, told with such candour, painted a picture of an era that I wholeheartedly hope to have authentically represented within these pages.

I would like to also thank my partner Michelle, who encouraged me throughout the many days and nights of writing this.

Finally, I would like to thank Dr Wendy Le-Las, Charles Pugh and Philip Mead, who were there for my grandmother when she needed them most. I feel very lucky to have met them all these years later and I am privileged to have heard their first-hand accounts.

APPENDIX

The Monkton Coke Works issue affected the lives of thousands of people in Hebburn, and inspired many to express the suffering that the plant caused for the community in various art forms, including painting, photography, and poetry. One notable poem is from Alastair Greason, who wrote the following in the year that Monkton Community Woodland and Business Park opened.

The Last Coke Works

Shrouding victims in living steam,
Watching over them, slowly draining hours from lives.
Jennie is right you know, time will prove it.
This burning Hell, on Monkton's Fell, stealing husbands,
kids and wives.

Two-faced giver of meagre wealth. Taker of rightful health.
Giant grey Ogre, stalking your neighbours, staring down
over their rooftops.
Never still, yet never moving,
And always awake.

You feed giant, mutant blackberries,
Sneaking rich poisons into our bodies,
Whilst sulphur coats the backs of throats.
Yet beauty lies in your sprawl,
Metallic music signals your life.

Nothing lies beyond your reach,
Until we run away.
Creeping into homes, shops, schools,
And the smallest of places;
Lungs, blood, cells, our very DNA.

Charging black ants, once green,
Amber eyes, glow high upon their heads.
Lead columns of blind, obedient red ants,
And feed your glowing belly with black breads.

You send these servants, to scuttle off,
Down George's, ancient, metal road.
Bearing your steaming fruits, warmth for the many, and
riches for a few.
Coating washing lines and window sills
With deceitful 'smokeless' gifts and ills.

Torrents pour down, subduing insatiable thirst.
Day into night, and back into day,
Then sending, white imposters to joyfully dance your thanks,
Billowing high into blue crystal sky,
Or coughing sideways, across the fields.

You can't hold back your anger,
Tentacles sweep, lash out, defile,
Gripping Hexham, Clyde, and Suffolk,
Before releasing, just for a while.

Mourning your sisters and brothers,
Their flames and bright arc lights turned off.
Revenge for Norwood, Hawthorn, Lambton,
Fishburn, Derwenthaugh?

But still your greatest might and wrath;
Ever unleashed upon little Lukes Lane.
Stunned, snared in your favourite path,
Perhaps because its folk know best, your pungent lies.
Now, tell them again, of smokeless joys, whilst another
widow cries.

— Alastair Greason, 2020

NOTES

Chapter One: Haw'way the Lads

1. Sutcliffe, Kay, 'A poem by a miner's wife,' *Velvetmedia,* May 8, 2015, https://velvetmedia. wordpress.com/2015/05/08/kay-sutcliffe-poem-by-a-miners-wife/.
2. Mason, Rob, 'Haw'way the Lads,' *Sunderland Association Football Club,* October 17, 2018, https://safc.com/news/club-news/2018/october/haway-the-lads/.
3. Ville, Simon, *Coal Was King of the Industrial Revolution, but Not Always the Path to a Modern Economy,* University of Wollongong, June 9, 2016, https://www.uow.edu.au/media/2016/coal-was-king-of-the-industrial-revolution-but-not-always-the-path-to-a-modern-economy.php.
4. Newcomen's engine was operated by condensing steam drawn into the cylinder, thereby creating a partial vacuum which allowed the atmospheric pressure to push the piston into the cylinder. It was the first practical device to harness steam to produce mechanical work.
5. Cooper, Tom, 'The meaning of a marra,' *The Northern Echo,* May 8, 2012, https://www. thenorthernecho.co.uk/opinion/letters/9693690. meaning-marra/.
6. Sweezy, Paul M., *Monopoly and Competition in the English Coal Trade,* Harvard University Press, 1938.
7. Knight, David, *Humphry Davy: Science and Power,* Cambridge, Cambridge University Press, 1992.
8. 'Monkton Coking Plant,' *Durham Mining Museum,* http://www.dmm.org.uk/colliery/m650.htm.
9. 'History of the Palmer Cranes at Jarrow,' *Tyne Built Ships,* http://www.tynebuiltships.co.uk/Palmer-Yard-Jarrow-Cranes.html.
10. Russell, William H., *The Deceit of the Gold Standard and of Gold Monetization,* American Classical College Press, 1982.
11. Constantine, Stephen, *Unemployment in Britain Between the Wars,* Longman, 1980.
12. Maconie, Stuart, *Long Road from Jarrow,* Ebury Press, 2018.
13. Ibid.
14. Ibid.
15. Perry, Matt, *The Jarrow Crusade: Protest and Legend,* University of Sunderland Press, 2005.
16. Ibid.
17. 'The Jarrow March,' *Trade Union Congress,* https://tuc150.tuc.org.uk/stories/the-jarrow-march/.
18. 'The Jarrow Crusade,' *UK Parliament,* https://www.parliament.uk/about/living-heritage/transformingsociety/electionsvoting/case-study-radical-politicians-in-the-north-east/introduction/about-the-case-study111111/.
19. 'Monkton Coke Works,' *Co-Curate,* https://co-curate.ncl.ac.uk/monkton-coking-works/.

Chapter Two: Home Sweet Home

20. 'Coal Industry Nationalisation Act 1946,' *The National Archives,* https://www.legislation. gov.uk/ukpga/Geo6/9-10/59/enacted/data. xht?wrap=true.
21. Hill, Alan, *The South Yorkshire Coalfield: A history and Development,* Tempus Publishing, 2001.
22. *Bad Neighbours - First Edition,* VHS, Tyne Tees Television, 1988.
23. 'Netty' is a Geordie term, used since the 1880s, to denote the word 'toilet', thought to originate from the Italian 'gabinetti', which means 'public restrooms'.
24. Alexander, Moira, Jennie Shearan's daughter, interview by author, Tyne and Wear, February, 2022.
25. Brown, Eddie, 'Cokeworks crusader's long battle,' *Newcastle Journal,* August 17, 1989.
26. Alexander, Moira, Jennie Shearan's daughter, interview by author, Tyne and Wear, February, 2022.
27. Traditional song.
28. Dandelion and Burdock is a beverage that has been consumed in the UK for centuries. It was originally a type of light mead, but has evolved into a carbonated soft drink commercially available today. https://www.fentimans.com/drinks/soft-drinks/dandelion-burdock.
29. 'Bairn' is a Geordie term to denote 'children', originating from the thirteenth century Old English word 'bearn'.

Chapter Three: Dante's Inferno

30. Shearan, Brian, Jennie Shearan's daughter, interview by author, Tyne and Wear, February, 2022.
31. Ibid.
32. Ibid.
33. Ibid.
34. Kelly, Terry, 'No coke … without fire!,' *Gazette*, January 17, 1992.
35. *The Land Trust*, https://thelandtrust.org.uk/space/monkton/.
36. 'Monkton Coke Works,' *Yorkshire Film Archive*, https://www.yfanefa.com/record/27354.
37. Interview by author, Tyne and Wear, February 2022.
38. Geordie term for Monkton Coke Works.
39. Rowland, John, 'Acid rain fumes belch a legal loophole,' Greenwatch, *Evening Chronicle*, 1989.
40. This was the tagline for Walter Willson's stores.
41. Jennie's later interview with *The Journal* would highlight just how close the houses were to Monkton Coke Works: 'Pollution probe at coke works,' *The Journal*, January 2, 1983.
42. Deane, Avril, 'Love affair with life,' *The Journal*, April 29, 1985.
43. 'Here's the choice when you vote,' *Gazette*, 1973.
44. 'Jenny takes a seat with the experts,' *Gazette*, May, 1973.
45. Cain, Bill, 'Battling Jenny attacks prices,' *Gazette*, 1973.
46. 'Get tough on prices' wives told by champion,' *Gazette*, 1973.
47. 'Ramp is smooth path now,' *Gazette*, 1973.
48. 'North gets watchdog on prices,' *Gazette*, 1973.
49. 'Public advice job keeps her home so busy,' *Gazette*, 1973.
50. The article entitled, 'Fight for TV fair deal,' *Gazette*, 1973, details Jennie's campaign for a fair deal. Rediffusion was a business that distributed radio and TV signals through wired relay networks. This was based on connecting homes with multiple twisted-pair cables. Each pair carried a single TV or radio channel. The system was provided in most UK towns in the 1970s. Selection of the TV or radio station was by means of a rotary switch, usually mounted on a wall or window frame close to the point of entry of the cable into the home. The TV sets used on this system were stripped-down TV sets with no tuner. Tenants in Hebburn were paying an extra 25p a week on their rents for piped Rediffusion, whether they had a Rediffusion TV set or not. Many tenants in Hebburn did not have a Rediffusion set and thus were paying money for nothing. Jennie met with the company to explain this and secured a deal with Rediffusion whereby the tenants could have adapters fitted to their homes to address this.
51. Brown, Eddie, 'Cokeworks crusader's long battle,' *Newcastle Journal*, August 17, 1989.
52. 'British Rail cake upsets councillor,' *Gazette*.
53. 'Jennie's place?,' *Gazette*.
54. 'Getting down to complaints,' *Gazette*.
55. 'Area action plan – plea for support,' *Gazette*.
56. Ibid.
57. 'Cinderella' area to be cleaned up,' *Gazette*.
58. Blenkinsop, Mike, 'Me granda was a pit yacker,' *Gazette*, 1977.
59. 'Street party is a huge success,' *Gazette*, 1977.
60. 'Margaret Thatcher: A life in quotes,' *The Guardian*, April 8, 2013, https://www.theguardian.com/politics/2013/apr/08/margaret-thatcher-quotes.
61. 'The wheels on the bus,' *The Economist*, September 28, 2006, https://www.economist.com/britain/2006/09/28/the-wheels-on-the-bus.
62. 'Did Margaret Thatcher really look down on women?,' *The Washington Post*, November 27, 2020, https://www.washingtonpost.com/history/2020/11/27/the-crown-margaret-thatcher-feminism-the-Queen/.
63. In her first Cabinet, 88% of ministers were former public school students, 71% were company directors and 14% were large landowners.
64. This was disclosed within the first public inquiry information pack.
65. 'Cokeworks plan attacked,' *Evening Chronicle*, 1989.
66. 'Fearful people give irate thumbs down,' *Gazette*, April 4, 1979.
67. Morris, Peter, 'Row – but plant gets go-ahead,' *Gazette*.

Chapter Four: Residents Unite

68. 'Battling Jenny attacks prices,' *Gazette*, 1973.
69. 'Jenny fights on,' *Topics Tonight*, 1980.
70. 'Cash hopes on pollution,' *Gazette*, September 6, 1982.
71. Cozens, Maureen, 'Dark shadow of the coke works,' *Sunday Sun*, August 31, 1986.
72. This information was disclosed in meeting minutes from National Smokeless Fuels obtained from The National Archives in Richmond, London.
73. 'Bag of dust aids coke works plea,' *The Journal*, August, 1982.
74. 'Rate cut 'win' in coke dust war,' *Gazette*, May 2, 1981.
75. 'Reduce our rates, residents demand,' *Gazette*.
76. 'Protesters fight for rates cut in Hebburn,'

Gazette, July 3, 1981.

77. 'Residents seek rates reduction,' *Gazette*, March 25, 1982.
78. 'Protesters fight for rates cut in Hebburn,' *Gazette*, July 3, 1981.
79. 'UK Economy under Mrs Thatcher 1979-1984,' *Economics Help*, March 30, 2007, https://econ.economicshelp.org/2007/03/uk-economy-under-mrs-thatcher-1979-1984.html.
80. 'Margaret Thatcher,' *The Guardian*, April 8, 2013, https://www.theguardian.com/politics/2013/apr/08/margaret-thatcher-quotes.
81. Letter from National Union of Mineworks to Tyne and Wear County Council, 1985.
82. Front-page headline, 'Battling Jenny!,' *Gazette*, April 24, 1985.
83. 'His hot shot Highness,' *The Journal*, May 3, 1985.
84. Deane, Avril, 'Love affair with life,' *The Journal*, April 29, 1985.
85. Ibid.
86. 'Hebburn served up on a plate,' Topics Tonight,

Gazette, March 10, 1986.

87. 'Metro station opens,' *Evening Chronicle*, March 19, 1986.
88. Halkerston, Judith, 'Every day is a bonus,' *Washington Times*, July 25, 1985.
89. Bailoni, Mark, 'The effects of Thatcherism in the urban North of England,' *Metro Politics*, https://metropolitics.org/The-effects-of-Thatcherism-in-the.html.
90. Cozens, Maureen, 'Dark shadow of the coke works,' *Sunday Sun*, August 31, 1986.
91. Stoner, Sarah, 'Choke Works – or improvement?,' *Gazette*.
92. 'Stop the yellow peril,' *Evening Chronicle*, June, 1986.
93. Letter from National Smokeless Fuels to The Chief Planner, Department of Planning, Borough of South Tyneside, February 11, 1987.
94. 'Borough of South Tyneside Town Development Sub-Committee,' *Planning Permission*, March 13, 1987.
95. 'Monkton's problems,' *Gazette*, July 1, 1989.

Chapter Five: Head to Head

96. Kelly, Terry, 'Residents demand: Clean up your act!,' *Gazette*, March 10, 1987.
97. 'Anger mounts over 'pollution',' *Gazette*, 1987.
98. 1,000 in coke 'filth' protest,' *Gazette*, March 12, 1987.
99. 'Action Group Film Pollution,' *Evening Chronicle*, December 10, 1987.
100. Letter from Kenneth Scott of Buckingham Palace to Jenny Lowry, November 25, 1987.
101. There remain many residents of Hebburn who remember this moment. One such resident is Mary Greenfield who, on 13 August 2019, recounted the chants on the *Pictorial Hebburn* forum on Facebook.
102. 'Placards out at cokeworks,' *Gazette*, May 5, 1987.
103. 'Residents in works demo,' *Gazette*, 1987.
104. '£400,000 Monkton clean-up plan fears,' *Gazette*, 1987.
105. Kelly, Terry, 'Councillor hits out on cokeworks plan,' *Gazette*, May 4, 1989.
106. Kelly, Terry, 'Group rejects report claims,' *Gazette*, 1987.
107. 'Action Group film pollution,' *Evening Chronicle*, December 10, 1987.
108. Interview with Roy Howson on *Look North*,

VHS, BBC, 1987.

109. J. Taylor, Managing Director of Coal Products Limited, wrote a public letter to the *Gazette* to explain this in detail.
110. The local Member of Parliament for Jarrow was also acutely aware of this issue and wrote letters to the manager of Monkton Coke Works to outline the issues highlighted by his constituents.
111. The memo stated, 'It is also important for Local Planning Authorities to seek advice before permitting development nearby, so as not to introduce an area which may be at risk development which caters for numbers of people; for example, housing estates.'
112. This information is present in Inspector Ridley's report from the public inquiry.
113. Extract from a letter written by a resident of Monkton Lane Estate.
114. Ibid.
115. Ibid.
116. Handwritten speech by Barbara Burns.
117. Kelly, Paul, 'Works fight will go on – residents,' *Gazette*, March 16, 1989.
118. Kelly, Terry, 'We've won our fight!,' *Gazette*, 1987.

Chapter Six: Elvis Has Left the Building

119. List of newspapers in the United Kingdom by circulation, *Wikipedia*, https://en.wikipedia.org/wiki/List_of_newspapers_in_the_United_Kingdom_by_circulation#1950-1999.
120. McLuhan, Marshall, *Understanding Media: The*

Extensions of Man, The MIT Press, 1994.

121. 'List of most watched television broadcasts in the United Kingdom,' https://en.wikipedia.org/wiki/List_of_most_watched_television_broadcasts_in_the_United_Kingdom

122. '25 years of Watchdog,' *BBC*, https://www.bbc.co.uk/pressoffice/pressreleases/stories/2005/01_january/07/watchdog.shtml.
123. Corr, Kate, 'BBC examines Monkton fears,' *Gazette*, 1988.
124. 'Watchdog,' VHS, *BBC*, 1988.
125. Ibid.
126. Ibid.
127. Ibid.
128. Ibid.
129. Ibid.
130. Ibid.
131. Monkton Coke Works meeting minutes from 1988, procured from The National Archives.
132. Kelly, Terry, 'Cokeworks probe delay is slammed,' *Gazette*.
133. Le-Las, Wendy, *Slaying the Dragon: The Demise of Monkton Cokeworks*, Le-Las Associates – Planning Consultancy, http://lelasplanningconsultancy.weebly.com/uploads/4/6/8/7/4687026/slaying_the_dragon.pdf.
134. 'Garden fete success,' *Gazette*, July 26, 1988.

135. 'TV probes poll,' *Gazette*, 1988.
136. 'Focus on Works,' *Gazette*, 1988.
137. 'Monkton in the spotlight at Side Gallery,' *Gazette*, 1988.
138. Alevroyiannis, John, 'Coke works: Now a call for medicals,' *South Tyneside Courier*, 1988.
139. 'Fears over Coke Works' fallout,' *Gazette*, 1988.
140. 'Guiding beacon for travellers,' *AFN*, November 10, 1988.
141. Bissett, J, 'Monkton cokeworks watched by night,' *Gazette*, 1988.
142. 'TV zooms in on Monkton,' *Gazette*, 1988.
143. 'Bad Neighbours - First Edition,' VHS, *Tyne Tees Television*, 1988.
144. Ibid.
145. Ibid.
146. 'Jennie's safety video,' *Gazette*.
147. 'We're not a rubbish dump,' *Evening Chronicle*.
148. A national magazine, *My Weekly*, featured Jennie and her story, and Jennie recounted this event during the interview. Nicholls, Margaret, 'The Dark Days Are Over,' *My Weekly*, 2001.

Chapter Seven: On the Road

149. Kelly, Terry, 'Clean-up pledge!,' *Gazette*, February 17, 1989.
150. Interview with Jennie Shearan, 'Look North,' VHS, *BBC*, 1989.
151. 'Wor Kate's cheque for Monkton action group,' *Gazette*, 1989.
152. 'Catherine Cookson sends money to help coke works campaigners,' *Gazette*, 1989.
153. Letter from Tom Cookson to Jennie Shearan, January 21, 1989.
154. 'Death in the Air,' *The Sunday Times*, 1989.
155. 'Cookson backs fight,' *South Tyneside Times*, 1989.
156. Kelly, Terry, 'Support pours in for the campaigners,' *Gazette*.
157. Front-page headline, 'Fight goes on,' *South Tyneside Courier*, February 9, 1989.
158. 'Coke works protest goes to Strasbourg,' *Gazette*, 1989.
159. Kelly, Terry, 'Cokeworks has no case to answer,' *Gazette*.
160. 'Commission backs bid to clean up cokeworks,' *Gazette*.
161. Dunn, Judith, 'Worthwhile pilgrimage to Europe,' *Gazette*, 1989.
162. Shipton, Martin, 'Action group adds fuel to its argument,' *The Northern Echo*, January 20, 1989.
163. 'Protest groups unite to fight pollution,' *Gazette*, March 14, 1989.
164. Allan, Bob, 'Protesters hit by a setback,' *Gazette*, March 18, 1989.
165. 'Cokeworks probe is doubted,' *Gazette*, June 30, 1989.

166. 'Coke works under fire,' *Gazette*.
167. 'Health leaflets handed out…,' *Gazette*, May 18, 1989
168. 'Leaflet probe into Monkton,' *Gazette*, May 17, 1989.
169. 'Health leaflets handed out,' *Gazette*, May 18, 1989.
170. Alevroyiannis, John, 'DIY health survey,' *South Tyneside Courier*, June 1, 1989.
171. Oldfield, Lesley, 'Health probe under way,' *Gazette*, May 24, 1989.
172. 'Doorstep survey reveals coke peril,' *Sunderland Echo*, July 3, 1989.
173. Hebburn Residents' Action Group, *Survey findings, Monkton Coke Works Area (Hebburn)*.
174. Hebburn Residents' Action Group, *Campaign Against Pollution*, 1989, distributed by Hebburn Residents' Action Group.
175. Kelly, Terry, 'Works are unhealthy says survey,' *Gazette*, July 3, 1989.
176. 'Campaigners fight for future,' *Gazette*.
177. 'Meeting at club on coke works,' *Gazette*, August 2, 1989.
178. Brown, Eddie, 'Cokeworks crusader's long battle,' *The Journal*, August 17, 1989.
179. 'Cokeworks pollution causes health problems – survey,' *Gazette*.
180. 'Support for battle on pollution at Hebburn,' *Gazette*, 1989.
181. 'Residents sit in at cokeworks plant,' *Gazette*, 1989.
182. 'Angry women stage sit-in,' *Gazette*, 1989.
183. Brown, Eddie, 'Coke works crusader's long

battler,' *The Journal*, August 1989.

184. Kelly, Terry, 'Hell-hole,' *Gazette*.

185. Kelly, Terry, 'Baby coughs up dust,' *Gazette*, March 16, 1989.

186. Kelly, Terry, 'Pollution fears grow,' *Gazette*, March 5, 1987.

187. 'Residents' fury,' *Gazette*.

188. Letter from Jack Douglass, 'Monkton: Enough is enough,' *Gazette*.

189. 'Being driven to the bottle by Monkton!,' *Gazette*, May 31, 1989.

190. Letter from John Badger, 'Tribute to residents is deserved,' *Gazette*.

191. Letter from Jack Wandlese to *Gazette*, 'Clean works, don't close it down,' 1989.

192. Letter from Margaret Landless to *Gazette*, 'Only residents know the truth,' August 11, 1989.

193. Kelly, Terry, 'Friends' leader accuses council,' *Gazette*, July 12, 1989.

194. Jamieson, Mike, 'What about people?,' Greenwatch, *Evening Chronicle*, March 21, 1989.

195. Letter from the Hebburn Residents' Action Group, 'Mike's report on coke works was accurate,' *Gazette*, 1989.

Chapter Eight: The White Knight

208. Kelly, Terry, 'No council cash for group case,' *Gazette*, January 5, 1990.

209. 'Legal aid lifeline,' *South Tyneside Times*, January 4, 1990.

210. Kelly, Terry, 'Cokeworks report goes to inquiry,' *Gazette*, 1990.

211. Report completed by Inspector Charnley, entitled *Re-opened inquiry into an Appeal by National Smokeless Fuels and Coal Products Limited*, with the dates of inquiry from January 30 to February 2 1990.

212. Brown, Eddie, 'Residents accused of scare tactics,' *The Journal*, February 1990.

213. Report completed by Inspector Charnley, entitled *Re-opened inquiry into an Appeal by National Smokeless Fuels and Coal Products Limited*, with the dates of inquiry from January 30 to February 2 1990.

214. Brown, Eddie, 'Research may still go ahead at plant,' *Gazette*, 1990.

215. 'Coke works: £50,000 survey is on the cards,' *South Tyneside Courier*.

216. Ord, Richard, 'Monkton: Two important meetings,' *Gazette*, July 1990.

217. Letter from Dr Suzanne Moffatt, Division of Community Medicine, to Jennie Shearan.

218. Brown, Eddie, 'Research may still go ahead at the plant,' *Gazette*, May, 1990.

196. 'A step forward,' *South Tyneside Courier*, July 27, 1989.

197. 'Monkton saga,' *Gazette*, 1989.

198. Taylor, Paul, 'Councillors join residents in new inquiry demand,' *Sunderland Echo*, June 22, 1989.

199. Corr, Kate, 'Probe into cokeworks is slammed,' *Gazette*, August 5, 1989.

200. Front-page headline, 'Cokeworks: New inquiry,' *Gazette*, August 4, 1989.

201. 'Council decision could re-open Cokeworks probe,' *Gazette*, May 5, 1989.

202. Front-page headline, 'Cokeworks: New inquiry,' *Gazette*, August 4, 1989.

203. Allan, Bob, 'Victory for residents,' *South Tyneside Times*, August 10, 1989.

204. Jamieson, Mike, 'How to help fight pollution,' Greenwatch, *Evening Chronicle*, December 19, 1989.

205. 'Legal aid lifeline,' *South Tyneside Times*, January 4, 1990.

206. Kelly, Terry, 'Cokeworks – start date for inquiry,' *Gazette*, 1989.

207. Brown, Eddie, 'Residents to oppose coke plant plan again,' *Gazette*.

219. Front-page headline, Kennedy, Kate, 'Cokeworks health probe,' *Gazette*, July 6, 1990.

220. Kelly, Terry, 'Cokeworks set to close down,' *Gazette*, September 22, 1990.

221. Front-page headline, Kelly, Paul, 'No reprieve for Monkton,' *Gazette*, September 29, 1990.

222. 'It's a case of no smoke without ire,' *Gazette*, 1991.

223. 'Jubilation at the Monkton news,' *Herald & Post*, 1990.

224. Le-Las, Wendy, interview by author, February, 2022.

225. 'Coke works to close,' *Evening Chronicle*, September 1990.

226. Kelly, Terry, 'Locals breathe a sigh of relief,' *Gazette*, December 29, 1990.

227. Front-page headline, 'Monkton only to be mothballed,' *Gazette*, October 23, 1990.

228. Jamieson, Mike, 'We rise from the muck,' *Evening Chronicle*, August 24, 1990.

229. Front-page headline, 'Survey is shunned,' *South Tyneside Herald & Post*, 1991.

230. Kelly, Paul, 'Response to survey is praised,' *Gazette*, 1991.

231. Front-page headline, Kennedy, Kate, 'Health survey into cokeworks,' *South Tyneside Times*, January 17, 1991.

Chapter Nine: Risking Everything

232. Front-page headline, 'Blot on landscape fear on cokeworks,' *Herald & Post*, 1991.

233. Mawson, Phil, 'Sensational announcement may not be end of matter,' *Gazette*, 1991.

234. Front-page headline by Mawson, Phil, 'Yes to power station,' *Gazette*, March 28, 1991.

235. Front-page headline by Ruane, Michelle, 'You're takin' the Michael!,' *South Tyneside Times*, April 4, 1991.

236. Covering the first three pages, 'Plant decision is condemned,' *Gazette*, March, 1991.

237. Flinn, Jon, 'Chimney pollution row goes to court,' *The Journal*, April, 1991.

238. Front-page headline, 'Appeal 2 – Action group's High Court bid,' *Gazette*, March, 1991.

239. Kelly, Terry, 'Cokeworks to be demolished,' *Gazette*, January 16, 1992.

240. Kelly, Paul, 'Health survey talks ban for protest head,' *Gazette*.

241. Kelly, Paul, 'Chokeworks,' *Gazette*.

242. There were subsequently several official studies and analyses on the Monkton Coke Works case.

One such study was led by Dr C. Miller from the University of Salford, who corresponded with Jennie about her request for several pieces of information.

243. Allan, Bob, 'Campaign for health takes step forward,' *Star Series*, March, 1992.

244. The article, 'Sense wins the day – in the end,' published by the *Gazette* on November 2, 1991, is an example of Jennie being invited to comment on local environmental issues outside of Monkton Coke Works. In this article, she commends the local council for 'at last seeing sense,' and abandoning a lake scheme.

245. 'Councillors get nod of approval from Jennie,' *Gazette*.

246. *Hebburn Focus*, ed. Bob Lambert, published by Councillor Keith Orrell on behalf of the Liberal Democrats.

247. 'Going with huge bang!,' *Gazette*, 1992.

248. Nicholls, Margaret, 'Woman with a mission in life,' *Evening Chronicle*, 1992.

249. 'Action group leader Jenny hits out,' *Gazette*.

Chapter Ten: Slaying the Dragon

250. Ord, Richard, 'End of grim landmark,' *The Journal*, November 5, 1993.

251. 'Look North,' VHS, *BBC*.

252. Kelly, Paul, 'Joint protest over cokeworks plan,' *Gazette*.

253. Ruane, Michael, 'Plant future undecided,' *Gazette*.

254. 'Leave us green!,' *Gazette*, August 10, 1995.

255. 'Cokeworks fire alert,' *Gazette*.

256. Front-page headline, Oliver, Gary, 'Yards from disaster,' *Gazette*, July 12, 1993.

257. McKay, Neil, 'Monkton may be used for new housing estate,' *Gazette*, January 18, 1992.

258. Kelly, Paul, 'Campaigners back cokeworks decision,' *Gazette*, November 15, 1993.

259. In Jennie's letter to Mr Pigg, Chief Planning Officer, she outlined numerous reasons why the heavily contaminated land needed to be carefully dealt with.

260. Howe, Jennifer, 'Toxic threat looms over park,' *Star Series*, February 27, 1992.

261. 'Call for a share of Euro grant,' *Gazette*.

262. 'Hope rises from the ashes,' *Gazette*, April 14, 1992.

263. Letter from Jennie Shearan to Mr Munroe, December 8, 1999.

264. Clark, T and Dilnot, A, 'Long-Term Trends in British Taxation and Spending,' IFS Briefing Note, June, 2002, http://www.ifs.org.uk/bns/bn25.pdf

265. 'Unemployed in Tyne and Wear,' *North Star Bulletin*, August 9, 2014, https://unemployed-

tynewear.wordpress.com/tag/jennie-shearan/.

266. Front-page headline, Kelly, Terry, 'Disaster team in cokeworks fight,' *Gazette*, October 31, 1990.

267. 'Residents in fight for compensation,' *Gazette*.

268. Jennie continued to write letters to key decision makers, just as she had throughout her campaign for clean air. There exists extensive correspondence between Jennie and the development control team of the Government Office for the North East on this matter.

269. Front-page headline, 'What price jobs?,' *Gazette*, July, 1995.

270. Jennie subsequently followed up with a letter detailing the lead up to the unfair headline, addressed to the senior editor.

271. Open letter from John Badger to the *Gazette*, 1995.

272. Letter from Jennie Shearan to Artist's Agency.

273. The extent of the contamination was detailed in a letter from J Ellis, the planning liaison officer to Mr Gray of South Tyneside Borough Council, December 21, 1995.

274. 'Monkton Community Woodland,' *The Land Trust*, https://thelandtrust.org.uk/space/monkton/#history.

275. This study was led by Artists' Agency, based in Sunderland, Tyne and Wear.

276. Kelly, Terry, 'Memories of our heritage,' *Gazette*, July 26, 1999.

277. 'Metal Guru forging Ahead,' *Evening Chronicle*, December 5, 2004, https://www.chroniclelive.

co.uk/whats-on/theatre-news/metal-guru-forging-ahead-1639571.

278. Transcription of Hilary's speech from family members of Jennie Shearan who attended the opening event.

Epilogue

281. World Commission On Environment and Development, *Our Common Future*, Oxford, Oxford University Press, 1987.

282. Pugh, Charles and Martyn Day, *Pollution & Personal Injury: Toxic Torts II*, London, Cameron May, 1994.

283. Miller, Christopher E., 'Planning, Pollution and Risk: Findings of Recent Research,' *The Town Planning Review* 65, no. 2 (1994): 127–41.

284. Pugh, Charles and Martyn Day, *Pollution & Personal Injury: Toxic Torts II*, London, Cameron May, 1994.

285. *Evening Chronicle*, January 21, 1992.

279. Kelly, Terry, 'Clean up for crusaders,' *Gazette*, June 26, 2000.

280. '20 Companies in NE31 2JZ, Monkton Business Park North, Hebburn,' *Endole*, https://suite.endole.co.uk/explorer/postcode/ne31-2jz.

286. Weiss, C, 'Women and environmental justice: A literature review. Women's health in the North,' *Gender and Disaster*, http://www.genderanddisaster.com.au/wp-content/uploads/2015/06/Doc-037-Women-and-EJ-Lit-Review.pdf, 2016.

287. The article, 'Show of strength,' written by Terry Kelly, and published by the *Gazette*, details the several scrapbooks that she and her daughter Barbara kept.

288. Rodgers, Richard and Oscar Hammerstein II, *You'll Never Walk Alone*, released by Gerry and the Pacemakers, 1963.

RESEARCH APPROACH AND SELECTED BIBLIOGRAPHY

My intention for this book has been to provide a highly accurate but also humanised account of Jennie's campaign for environmental justice in her community. I have endeavoured to be rigorous in my research.

The below list of books, official reports, and academic papers have provided me with a significant amount of information. This bibliography is by no means a complete record of all the works and sources I have consulted. It indicates the substance and range of reading upon which I have formed my ideas.

In addition to the below reading, I have searched through considerable footage that was stored on VHS, newspaper clippings that were chronologically ordered in scrapbooks, reports from both public inquiries, and carefully catalogued letters to and from the Hebburn Residents' Action Group. Furthermore, I conducted interviews with several people from the area, and with those who were closely involved in the case.

Bew, John, *Clement Attlee: The Man Who Made Modern Britain*, Oxford, Oxford University Press, 2017.

Bhopal, Rajinder S., Peter Phillimore, Suzanne Moffatt and Christopher Foy, 'Is Living near a Coking Works Harmful to Health? A Study of Industrial Air Pollution', *Journal of Epidemiology and Community Health* 48, no. 3 (1994): 237–47.

Bourn, John, *Control and Monitoring of Pollution: Review of the Pollution Inspectorate*, National Audit Office, 1991.

Foot, Michael, *Aneurin Bevan: A Biography: Volume 1: 1897–1945*, Faber and Faber, October 2009.

Jones, Erik, *The Oxford Handbook of the European Union*, OUP Oxford, 2014.

Le-Las, Wendy, *Understanding the Development Jigsaw*, Wiltshire, Buccaneer Books, 1997.

Le-Las, Wendy, *Slaying the Dragon: The Demise of Monkton Cokeworks*, Le-Las Associates – Planning Consultancy. http://lelasplanningconsultancy.weebly.com/uploads/4/6/8/7/4687026/slaying_the_dragon.pdf.

Miller, Christopher E., 'Planning, Pollution and Risk: Findings of Recent Research', *The Town Planning Review* 65, no. 2 (1994): 127–41.

Moore, Charles, *Margaret Thatcher: The Authorized Biography: Volume I: From Grantham to the Falklands*, Vintage, 2015.

Paxman, Jeremy, *Black Gold: The History of How Coal Made Britain,* William Collins, 2021.

Perry, Matt, *The Jarrow Crusade: Protest and Legend*, University of Sunderland Press, 2005.

Pugh, Charles and Martyn Day, *Pollution & Personal Injury: Toxic Torts II,* London, Cameron May, 1994.

World Commission On Environment and Development, *Our Common Future*, Oxford, Oxford University Press, 1987.

INDEX

ABOUT THE AUTHOR

Photograph by Victoria Rosolia

Gianfranco Rosolia is a Literature graduate from the University of Cambridge and has an MBA from Bayes Business School. *CLEAN AIR* is his first book. Born and raised in the North East of England, he currently lives in Los Angeles with his partner Michelle.